The Door to NowHere

A Portal into the Timeless Reality of Non-dual Consciousness

By ONCE WAS A SEEKER

This book is a work of non-fiction.

A catalogue record of this book is available from
the National Library of Australia

Cover art formulated with the use of ChatGPT

Disclaimer

The views expressed are those of the author and are offered for informational and contemplative purposes only. The author and publisher assume no responsibility for any consequences arising from the interpretation or application of the material presented.

For permissions, inquiries, or rights requests, please contact:
oncewasaseeker@gmail.com

First Edition
Paperback ISBN: 978-1-7641875-0-3
Published by: World2-0 Publishing

DEDICATION

To my darling wife, my partner of over 50 years,
whose love and patience have been unwavering.
(well, mostly—except when I've behaved badly). ☺

To my children and grandchildren, whom I love—
and who have loved me in return.

You've long tolerated my 'worldview' with a wry smile,
especially when the subject of death arises—
knowing full well that I don't believe Life can die.

With deep gratitude for your presence in my life.

CONTENTS

PREFACE

Whether you come with a mind steeped in science, faith, or simple curiosity, this book has found you.

It explores the fundamental nature of *being* through the lenses of science, introspection, and religious revelation.

This preface includes two sections on the use of words. You are encouraged not to skip them, as they are essential to how the book is read.

During this journey, you may find the answer for yourself to that most fundamental question:

Who am I?

- A Being of flesh and blood?
- An Illusion created by your brain?
- A Mind?
- A Soul or Spirit?
- A Phantom in a dream?
- Or no *thing* at all?
- Perhaps, *some*, or *all*… and yet none.

***

Most stories are written to entertain.
Some aim to explain the world.

A few try to enlighten—inviting the reader to stop thinking and simply *be*. (But how could you not *be*, even while thinking?)

In an absolute sense, this story is of the first kind: pure entertainment. For it cannot change who you *Really* are.

Which is… for you to discover ☺

This book simply invites you to look deeply into the facets of your being.

The Significance of Words

As this book is full of words, it's worth taking a moment to consider how they are used.

Words as Ideas

Throughout these pages, you'll find the words *Consciousness and Awareness,* as well as *God*, *Allah*, *Krishna*, *Suchness*—all referencing *Ultimate Reality*.

Such words often carry strong associations with *ideas*.

- The *Atheist* rejects an *idea* of God.
- Many *Theists* believe in an *idea* of God.
- Many *Buddhists* cling to the *idea* of Enlightenment.
- Many *Hindus* feel bound by the *idea* of Karma.
- Many *Scientists* hold to an *idea* of Reality.

Yet no *idea*[1] can grasp *Reality*.

Reality is literally inconceivable—as will become clear when we explore the nature of *Knowing-ideas*.

***

Reading these words puts the '*key of attention'* in the lock of the *Door to NowHere.*

Thinking about their *meaning* (*ideas*) turns the key… but opens the Door back into the world of dreams.

So how can a book of words say anything about *Reality* at all?

Fortunately, there is a way out of this seeming impasse.

[1] Try to define any idea without referencing other ideas. It cannot be done. And those ideas reference other ideas, and so on. It's impossible to use ideas to get beyond ideas to *Reality* itself.

Words as 'Pointers'

While most words are associated with *ideas*, there are a few that, when *stripped of meaning*, become clear pointers to *Reality*.

Take the words '*Seeing-colours*'.

You can read countless scientific theories *(ideas)* about *Seeing* and *colour*, but none reveal the *experience* itself.

In simply ***looking****, the Reality referenced by the words 'Seeing-colours' is plain*

The words *Seeing-colours* are also hyphenated for a reason.

Pause and consider:
No *colour* can exist without the capacity to *See.*

Just imagine what your elbow *perceives* to recognise that *colour* and *Seeing* are inseparable: 'non-dual'.

Similarly, if you reflect for a moment, it is *self-evident* that the words *Seeing-colours—along with Hearing-sounds, Tasting-flavours, Smelling-odours, Feeling-feelings, and Knowing-ideas*—all point to ***non-dual facets of Awareness****.*

'Non-dual' simply means *Awareness* cannot be divided into parts—except conceptually.

Just as a coin may be seen as having two sides, yet is made of a single undivided substance, so *Awareness* may appear to have many facets—yet remains whole.

That is why you'll find these words capitalised, hyphenated, and italicised throughout the book—to highlight what is *self-evident…* to *you* (when *you* pay attention to where the words 'point').

Some may prefer *Consciousness* over *Awareness* or other terms altogether—*God, Truth, One, Love, Divine, Suchness.* All these terms are used in the book to resonate with different worldviews about *Ultimate Reality.*

Rather than debate the words, simply **notice** what they point to in your own *experience*—given their context in the book.

***

The words *material* and *immaterial* are also italicised to highlight a distinction that is usually not evident in daily life.

An Invitation to Attend

With so much to read and watch, it's common nowadays to skim, or to now use AI to "get the gist." But here, that won't do.

This book intentionally uses several devices to interrupt the flow of *thought*, to allow *attention* to turn 'inward'.

Many words are italicised or bolded. Sentences are often broken into short segments, with frequent headings and space between sections. I use ellipsis and dashes… a lot ☺

Throughout, you'll also see the words "please do it," inviting you to undertake some experiential exercises, along with words like *self-evident*, *plainly*, *notice*, and similar phrases. These are not to impose *belief*—quite the opposite.

All these 'devices' simply invite you to *not* take the writer's words as *Truth*.

Instead, pause, and *look for yourself*. Is it '*self-evident*' to *you*?

And, if you find yourself baulking at some word. Consider: are you associating the word with an *idea*, instead of *attending* to the *Reality* it is referencing?

As an example, '*void*' could be such a word, which is later explored *experientially*.

A Word of Caution and A Word of Hope

Saint Augustine once warned:

> *"We must be on guard against interpretations of Scripture that are too far-fetched or opposed to science, and so exposing the word of God to the ridicule of unbelievers."*

This book does not oppose reason, deny science, or dismiss faith. Rather, it shines a light on the deeper wisdom behind all three—by exploring the science, undertaking introspection, and looking to where all the great religions point.

And perhaps, surprisingly, there is a convergence to discover.

***

If you have a deep relationship with God—whether through Christianity, Islam, Judaism, Hinduism, or another path—this book does not ask you to let go of that devotion.

Instead, it invites you to see what that relationship already is.

The *Divine* is not far off, waiting to be reached. It is not something to merely believe in or worship from a distance. It is *NowHere.*

The mystics of every tradition have spoken of this closeness:

- **Christianity:** "Christ is in you."
  — (Colossians 1:27)

- **Islam:** "We are closer to him than his jugular vein."
  — (Quran 50:16)

- **Judaism:** "Where can I flee from Your presence?"
  — (Psalm 139:7)

- **Hinduism:** "I am the Self... seated in the hearts of all beings."
  — (Bhagavad Gita 10:20)

Or as Meister Eckhart[2] said:

"The eye through which I see God is the same eye through which God sees me."

If you read with an open heart and a still mind, you may find that the One you've always loved was never separate. Not for a moment, through all your sorrows and joys.

Instead of Love being lost in these words… it may be that Love itself blossoms.

***

And if you are a Buddhist, a scientist, or simply a sceptic, this book too will dig deep into your own *beliefs*—to separate *idea* from *Reality*.

***

Fundamentally, this book challenges neither science (which evidently works), nor religion (whichever you are drawn to), but…

Your own *assumptions* about the *Ground* of both.

What Do You Get Out of It?

Perhaps, at some point, you'll get a good laugh.
The kind of laugh that arises in the final stanza of the *Mini-Mahābhārata*, a fable about *Reality* at the end of the book.

Along the way, it may challenge everything you take for granted.
Or it may simply restate what is already plain to you—another joyous affirmation of *Life*, just as it is.

---

[2] A 13th Century Christian mystic. https://en.wikipedia.org/wiki/Meister_Eckhart

The Red and Blue Pills

For fans of *The Matrix*, continuing to read is equivalent to taking the 'red pill'—but only if *attention* is directed to the *Reality* being referenced.

Getting stuck in *thought* (*ideas*) is the same as taking the 'blue pill'. ☹

On some accounts, 'taking the red pill' is shorthand for 'becoming spiritually enlightened'— realizing the true nature of '*Reality*'.

The Bad News

No person ever has, or ever can, 'become enlightened'…

Any more than a character in a film can step off the screen

This is not just a claim to be believed or disbelieved. Rather…

Some Good News

It's *directly verifiable* via an *enlightening experience.*

Which raises the paradox… if no person can ever become enlightened, how can an 'enlightening experience' occur?

That, my friend, is the billion-dollar koan[3] that this book invites you to explore.

But first…

More Bad News

The *experience* of 'enlightenment' (if it arises) arises *spontaneously*. Which is to say: there is no *path* or *practice* that will trigger it.

---

[3] Allowing for inflation of the 'sixty-four-thousand-dollar question'

Not even reading this book. ☹
So, why bother writing it—or reading it?

Here is why…

The Book is Chekhov's Gun

Chekhov's Gun is a principle from theatre that states:
If a gun appears in the first act, it must be fired by the third.
In short, every element in a story should matter.

This book is one such element in the unfolding plot of our life stories.

Just as it's been my apparent destiny to write it,
it seems it is yours to read it (at least to this point). ☺

Reading it may seem to bring a shift in outlook.
But like a gun fired on stage…
It holds no *Real* or lasting consequence.

Perhaps 'the person you seem to be' will be 'killed' by an 'enlightening *experience*'. Which may appear deeply consequential… until it happens.

Or maybe nothing will change, depending on how your story unfolds.

But you'll never know unless you read on…

Even More Bad News

Like all *experiences*, 'enlightenment' comes and goes. It never lasts.

What appears to be a destination is more like realising in a lucid dream that you are dreaming…

It is always followed by another dream.
Some dreams are lucid, most are not.

Same with this 'dream of life'… constructed by your thoughts.
Occasionally it is realised '*This is a Dream*'… mostly not.

PREFACE

The Very Good News

Reality is not a dream. It is always *NowHere.*

It does not matter which pill is chosen.
Or what the apparent effect is or is not.

Reality is in no need of enlightening…
And there is only Reality.

The door to *Reality* is already unlocked—and wide open.

This book is merely one apparent portal that disappears as soon as it is passed through… like a door in a dream.

There is no prize at the end. No final enlightenment badge.

Reading this book is like playing a game or dancing, or singing, without getting anywhere—simply for the joy of it.

It may just be the most wondrous journey you never expected…

The Road Home

The Journey Starts *NowHere*

Awareness *Is*

Plainly, you're *NowHere… Aware.*
This is not a belief, nor does it require a theory to be validated.

You take it for granted.

All exploration of *Reality* must start *NowHere.*

What you're *Aware* of is your body, thoughts, feelings, and surroundings. These are *objects of perception*—and they, too, are usually taken for granted.

Together, they form 'the person' you appear to be.

We rarely think of surroundings as part of 'who I am'. Yet they are.

A firefighter includes fire.
An actor includes audiences.
A hunter includes the forest and the field.

In Part Three: Introspection you'll get to examine this 'person' in much more detail, to see how they relate to *Awareness*; and how *Awareness* relates to *Consciousness.*

For the remainder of the book, the focus will be on *Consciousness* as the *Ground*, with *Awareness* an 'aspect' of it[4].

The importance of this distinction will become evident as we explore the *Power* that *manifests* the *Perceived World* within and to *Awareness.*

Next is a conceptual map and a broad synopsis of the journey ahead.

---

[4] These terms are often conflated. Also, some people may view *Awareness* as the *Ground* of *Consciousness.* The point is not to debate the terms, but to look at the *Reality* being referenced.

Waypoints on The Journey

The Limits of Science

1. The Paradox of *Material Perception*
2. Science and *The Objective Universe*
3. Science and *Consciousness*

Recognizing Reality

4. Looking for *Consciousness*
5. Finding *The Single Eye*
6. Perception is *Illusion*
7. The Nature of *Sensation*
8. The Nature of *Idea*
9. The Nature of *Power*

Consciousness and *Illusion*

10. Splitting *Non-dual Consciousness* in Two
11. The *Appearance* of *Change* and *Time*
12. *Duality* and *The Veil of Ignorance*
13. *One Life*, Many *Appearances*
14. *Apparent Life* is as *Real* as it Gets
15. *The Illusory Universe* is *Perfect*

Consciousness and Religion

16. Religious *Revelation*

Consciousness is *Reality*

17. *Consciousness* Alone *Is*
18. *This is Thou*

Let the journey begin…

The Limits of Science

1. The Paradox of *Material Perception*

Image courtesy of ChatGPT

If you begin with the assumption that the *apparent universe* is *material* and then trace the processes of *perception* (as theorised by neuroscience), you soon arrive at a paradox.

Take *Hearing-sounds* as an example.

On the *material* view, *material* sound waves hit the *material* eardrum, convert to *material* electro-chemical impulses in the inner ear, that then travel through the *material* auditory nerves to the *material* auditory cortex, which is a part of the *material* brain inside the *material* head.

These impulses are then 'somehow' converted (no one knows how) into the *experience* of *Hearing-sounds*—where the *sounds*, and *Awareness of the sounds*, arise *together* as a *fleeting immaterial* phenomenon.

Similarly, with *Seeing-colours.*

Light is said to be *material* electro-magnetic waves that are converted in the retina to *material* electro-chemical impulses that travel through the *material* optic nerves. Once these impulses enter the *material* visual cortex at the back of the *material* brain (in complete darkness)…

The impulses are also said to be converted (no one knows how) into an *immaterial experience*: *Seeing-colour*, where…

The colours themselves are the illumination.

Similarly for *Feeling-feelings, Smelling-odours*, and *Tasting-flavours.* All appear inside the brain as *inseparable immaterial* phenomena.

***

In *material* theory, there is no 'little person' inside the head looking out through the eyes into the *assumed material* universe.

The world that is *perceived* is composed *entirely* of *immaterial colours, sounds, odours, flavours, and feelings.*

As we'll explore, these *sensations appear* to be 'things, properties, processes, events, and relations', *only* as specific *immaterial ideas* [5] are associated with the fleeting *immaterial sensory images.*

This is not how most people view themselves and the world!

***

Common understanding is more like the main picture at the start of this section. It portrays a 'world of matter', including people with their own body and head.

***

The inset illustrates the 'first person perspective', which is very different to common understanding.

From this perspective, it is normal to *See* only *bits* of your own *apparent* body (like an arm and hand), while also *experiencing immaterial feelings* and *thoughts.*

You can verify this by simply checking out what you can *See* right *Now*—though we'll shortly do a couple of experiments to clarify the *Reality of what it is that you See, when you look.*

---

[5] As we will further discuss, *ideas* also have three 'facets': *number* (e.g. 'one'), *form* (e.g. 'spherical'), and *meaning* (e.g. 'ball').

Regardless, the science demonstrates:

All appearances are immaterial and inseparable from the immaterial Awareness in which and to which they appear.

This is the Paradox of *Material Perception*:

The universe *appears* to be *material*,
and science treats it as *material*,

but based on the *material hypothesis*,
the *Perceived Universe* cannot be *material*.

And it is only the *Perceived Universe* that can ever be *directly observed*, where:

Observer and observed are One:
Immaterial and Non-dual.

On this theory, no person (including any scientist) has ever *Seen* or *Felt* or *Tasted* or *Smelled* or *Heard* or *Known* anything made of *theorised* matter. Nor can they (or you)… Ever.

So where does this **Paradox** leave 'science'?

2. Science and *The Objective Universe*

Image courtesy of ChatGPT

The Role of Science

Science does an extraordinary job of *modelling* the *Perceived Universe.*

Its most rigorous *models* use mathematics to describe ***theoretical***:

- **Forms** – from the quantum field to cells, stars, and galaxies
- **Properties** – such as mass, charge, spin, and more
- **Constants** – like the gravitational constant
- **Laws** – such as conservation of momentum

A *theory* is considered valid when its *predictions reliably map* onto *perceived phenomena*—an *apparent*: falling apple, a microscope image, or telescope data.

These theories offer a glimpse into the astonishing depth and coherence of the 'objective universe', as well as gifting us with remarkable technologies.

The Objective Universe

Given the **Paradox**—which reveals that the *Perceived World* is *immaterial* and *inseparable* from the *Awareness* in which and to which it *appears*…

How can we account for the so-called 'objective universe'?
The universe that seems to stretch across all scales of space and time—that appears distinct from 'me', yet somehow includes me?

Since the *material* universe is only a *supposition*—one that can never be confirmed—we may simply dispense with it.

The *Ideal* Objective Universe

Instead, think of the 'objective universe' as more like *software* than *hardware.*

Like the software that determines the features of a computer game.

(This is just an analogy. Not *Reality*)

In this book, we'll refer to this 'software' as the *Ideal Objective Universe* (or simply the *Ideal Universe*). It contains all the *rules* governing the *form, behaviour* and *meaning* of *appearances.*

What this vast *Ideal Universe Is* 'in itself' is impossible to *Know.* Like the *supposed physical universe*, its 'essence' remains hidden.

What is *perceived* are narrow representations of it—filtered through the lens of each *being*, in a particular place, at a particular time, looking in a particular direction.

It's like viewing a computer game from the viewpoint of a character—where the images on the screen are nothing like the game's software.

Which leads to another question we'll look at later:
What is the 'underlying hardware' that supports this 'software'?

For now, it suffices to recognise that science gives some insight into the rules of the 'Objective Universe' (whatever it's basis).

Next, we look at what science cannot reveal.

3. Science and *Consciousness*

Image courtesy of ChatGPT

While the scope of science is immense, it is limited to the *Perceived Universe.*

As the **Paradox** reveals, and we'll further investigate, *perceptions* are made up of:

- *Immaterial sensory images* – composed of *colours, odours, flavours, feelings, and sounds*, and
- *Immaterial ideas* – of *number, form, and meaning*

Which *together appear in* and *to immaterial Awareness* as 'things, properties, processes, events, and relations', that are *apparently* separated into '*me'* and '*the world'*.

No matter how 'solid' and 'real' *apparent* things seem, based on the **Paradox**, their 'essence' is necessarily *immaterial…* like a dream.

This means that *every scientific instrument*—and *every measurement*—is itself an *immaterial perception.*

This is not to deny neuroscience. Clearly, it demonstrates a strong correlation between '*perceived* brain activity' and *Perception…*

But *correlation is not causation. A matter we'll return to later.*

***

Regardless, the **Paradox** means that no science can ever discover the *essence* or *true nature* of *sensations* and *ideas*—because it is not possible to look past the *immaterial perceived instruments and observations* to *the underlying Reality.*

Nor can any *formula* ever capture the *essence* of *lived experience…*

The hush of dawn over the desert,
the thrill and terror of a fight to the death,
the hug of a child,
the visceral disgust at the stench of rotting garbage,
the swelling notes of Mozart,
or the driving beat of an African drum...

Or anything else that makes *apparent* life what it *Is.*

Nor can *Awareness* or *Consciousness* itself be the 'object' of scientific study, for the simple reason:

Awareness is This in which, and to which,
all theories, instruments, and observations appear.

For *Now*, simply *be Aware* that *Awareness* (and its *Ground*, *Consciousness*) cannot be captured by any *theory,* simply because it is not an *idea.*

It is *Real…*

Not something that can be defined—
Only *Recognised.*

Recognizing Reality

4. Looking for *Consciousness*

Image courtesy of ChatGPT

So, let's begin. Not with belief or speculation.
But with the *recognition* that *Consciousness is NowHere.*

Perhaps the *idea* still lingers in *Awareness* that *Consciousness* is a function of the brain inside the head.

If so, let's test the *theory*—like any good scientist would.

Place one hand on each side of your *apparent* head. (Please do it, don't just think about it. Thinking just puts *attention* on '*ideas*'.)

Notice: the *feeling* is in *Awareness*,
along with the *idea*: *"My hands are holding my head."*

Now **notice**…

Neither the feeling, nor the idea, is an actual head. And any image of a head Seen in a mirror is likewise just that—an image.

Based on the **Paradox**, it's evident…

The 'head'—whether Seen, Felt, or thought about—is an immaterial appearance in and to immaterial Awareness that is One with Consciousness.

And since, from the 'first person' perspective, the *apparent* head contains the *assumed* brain…

Yet *Consciousness contains the apparent head,* it follows that:

The assumed brain cannot contain Consciousness.

As for any other *apparent* being's brain… that too is merely a *perception* in *Consciousness.*

Continuing the software analogy: 'bodies' and 'brains' are more like *subroutines in The Ideal Universe,* than 'physical things'.

Of course, *thoughts* may still assert: *"My real, material brain must exist."*

But this can never be proven.

For, as the **Paradox** reveals, there is no way to get beyond *appearances* to *observe* the *assumed material* universe.

As noted earlier, this in no way invalidates the findings of science—*as they apply to the Perceived Universe.*

It simply eliminates the need for the material assumption.

---

5. Finding *The Single Eye*

Image courtesy of ChatGPT

So, if *Consciousness* is not in any 'brain'. Where is it?

Look straight ahead. Without moving the gaze (please do it).

Notice the *visual field* appears as a circle with an indefinite edge.

Notice that 'beyond this edge' (surrounding the whole *visual field*) is neither 'black' nor 'white'…

Simply *Void.*
(Shown in the image above, and the image of the head and mirror, as the surrounding 'grey').

This is a 'pointing' word. It cannot be understood. It can only be directly noticed.

Just take a moment.
Be sure the '*Void*' is *Recognized.*

Now, **Notice**: no 'body' is *apparent.*

Now look down.
Notice: only the front of the torso and limbs *appear*—but no head.

Where the head should be, the whole world *appears*
(or at least the portion *perceivable* from the perspective of the person *One appears* to be).

***

This *visual field* is the *Eye of Consciousness.*
Some call it the *Single Eye of God.*
It is the *only Eye* there is.

It's *apparent* that this *Eye* is in no body—for all bodies (including the one *you* call 'my body') are *appearances* in and to it.

This *Eye* not only *Sees*, but it also *Hears*, and *Feels*, and *Tastes*, and *Smells*, and *Knows.*

This *Consciousness* is not in space or time, for 'space and time' are also *appearances* in and to *It.*

It is simply *NowHere.* Like a luminous *immaterial* jewel, suspended in the *Void.*

***

Now that your *Eye* is fully opened to the whole visual field…

As you go about 'daily life', simply notice from time to time the complete field of view—to the edge of the *Void*—and recognise that this *wonder full* 'play of life' is present within it. *NowHere.*

6. Perception is *Illusion*

Image courtesy of ChatGPT

An *illusion* is something that *appears* to be other than it is.

While the meaning of the word is easily understood,
the *Reality* becomes plain only through *direct observation* of the *Illusion* being pointed to.

Try this:
Hold up your thumb.

Notice: it *appears* larger than things you *Know* to be bigger.
But a thing cannot be both larger and smaller than another thing.
That it *appears* so, is clearly an *illusion.*

And what is true of the *apparent* 'thumb'…
is true of all *apparent* things.

***

The *illusion* runs even deeper.
For in *direct experience*, no 'things' *appear* at all—only *sensations.*

7. The Nature of *Sensation*

Look closely at any colour. Say, the 'yellow box' above, or a plain sheet of coloured paper, or a blank wall.

Observation reveals that the *colour* itself has no intrinsic *meaning.*

Logically, no combination of *meaningless colours* can give rise to *meaningful appearances*. Which is to say:

No colour, or combination of colours, can become a 'thing'.

The *true nature* of this *Seeing-colours* cannot be said.
For all saying is based on *ideas*—and *Seeing-colours* is not an *idea.*
It simply *Is.*

***

What is true of *Seeing-colours*, is true of *Hearing-sounds*, *Smelling-odours*, *Tasting-flavours*, and *Feeling-feelings.*

None has any form, or substance, or *meaning*.
Each is *immaterial*.

You cannot 'touch' a *sensation*. For, by the **Paradox**, the very hand that reaches is itself a *sensory image*.

One *immaterial sensory image* cannot touch another *immaterial sensory image*, any more than an image of a car in a film touches the image of the road. They both appear in the same 'field of view'.

Through *direct observation* it is *plain* that:

Sensations are meaningless and immaterial...
Inseparable from Awareness of them.

So where does *meaning* come from?

How is it that I *appear* to *Sense* 'things'—and not merely *sensations*?

The answer is also *plain*… if you '*think*' about it ☺

8. The Nature of *Idea*

It is *ideas* that give *sensory images* their apparent *number*, *form*, and *meaning*.

(For the sake of brevity, '*number, form, and meaning*' is sometimes shortened to simply '*meaning*', unless the context requires otherwise).

Ideas cannot be *Sensed*, for they have *no colour, odour, flavour, sound, or feeling*.

Yet *Knowing-ideas* is *self-evident*:

Two (number) cubed (form) dice (meaning).

It's also *self-evident* that *ideas* are not 'in' any words (written or spoken). Otherwise:

- one word couldn't have many *meanings*,
- nor could many different words (including in different languages) point to the same *idea*.

It is also clear that the *same idea* may be associated with *vastly different sensory images*, not just words.

The *same idea* can give *meaning* to:

- the *apparent* object itself,
- an *apparent* picture of the object,
- an *apparent* hallucination or dream about the object,
- or an *imagined* object…

The only *apparent* difference is in the other *sensations* and *ideas* that form the context.

For example, the *idea* of 'mother' may apply to my mother, your mother, or even a lioness with her cubs.

***

Or take the classic *illusion* of the 'old-young woman' that heads this section.

When the *idea* of 'a young woman looking away' is associated with the lines, that is what appears.

Now **notice**—without any change in the image—as the *idea* of 'an old woman with a large nose and chin looking down' arises in *Awareness*, that is what the *image appears* to be.

***

But the drawing is really *neither.*
It's just a set of squiggly lines.

That it *appears* to be anything at all is an *illusion*—the result of *ideas* 'projected' onto shapes.

Crucially, the *ideas* that provide *meaning* are not necessarily verbalised—though they can be.

You don't have to say to yourself what the image is.

You simply *See* the *image* as a specific object as the *idea* arises unspoken in *Awareness.*

Conversely, until a specific set of *ideas* is associated with a specific sensory image, there is no *Awareness* of any 'thing'. This experience will be examined in depth in Part Three: Introspection. It is usually *experienced* when clearly *Seeing* 'something' but not *Knowing* 'what' it is, until a specific *idea* is associated with the image.

***

There is no way to *explain* how the association of *ideas* with *sensations* happens—for every explanation relies on *ideas*, and the 'how' is not itself an *idea.* It is a function of *Consciousness* that can never be explained.

***

It's also self-evident that:

- No *idea* can exist apart from *the capacity to Know*, which is a *facet of Consciousness.*
- No *idea* can ever change. If it did, it would no longer be the *idea* that it is.
- Every *idea* takes its *meaning* only by reference to other *ideas.* These, in turn, refer to still more *ideas*[6] that, in the ultimate, form the closed set of *all possible things, properties, processes, events, and relations.*

It **necessarily** follows that *Consciousness* is *All-Knowing*

In religious language, *this Knowing-all-ideas*—unchanging and entire—is called *Omniscience.*

It includes the *idea* of the entire Universe: *The Ideal Universe,* which we will explore in depth in Part Two: When Science Meets Awareness, and in Part Three: Introspection.

---

[6] Try to define any *idea* without reference to any others—you'll find it cannot be done.

THE DOOR TO NOWHERE

This *All-Knowing* cannot be grasped. For it is not an *idea.*
It is *Real.*
It is *Conception* itself.

What is *Conception*?
What is an *idea*?...

No *idea*! ☺

It's not possible to conceive of 'Conception'.
It's like trying to *See* you own eyes with your own eyes.
It cannot be done.

Yet, *Conception* is *self-evident*: you *Know* that you *Know-ideas*…

At least *some ideas.*

It's not at all clear that you *Know-all-ideas, ever-unchanged!*
You'll get the chance to examine this 'veil of ignorance' shortly.

***

First, we need to address *Power.*

How are *sensations* and *ideas* fashioned into the *appearance* of 'things, properties, processes, events, and relations', so…

Some appear in and to Awareness as 'me' (bits of my body, my thoughts, and my feelings), and *some appear as* 'the world' (my surroundings)?

9. The Nature of *Power*

Image courtesy of ChatGPT

Power manifests the *apparent* universe in and to *Awareness* as patterns of *immaterial sensations* that accord with the *immaterial Ideal Universe.*

Together, these *sensations* and *ideas* create the *appearance of a solid, 3D world*—observed from the 'first person' perspective of an *apparent being*…

As if looking out of its eyes, and experiencing its body, thoughts, feelings, and surroundings as: '*my body, thoughts, feelings, and surroundings*'…

As in a dream.

This 'first person' view is illustrated in *Finding the Single Eye*, and the image above.

***

This *Power* is not merely 'raw energy'.
It *inherently Senses-sensations* and *Knows-all-ideas.*

Without *sensations* and *ideas* arising in harmony, there could be no *appearance* of a *coherent, meaningful, intelligible world*—nor any 'first person' experience.

It would be nothing but incoherent 'noise'.

As dreams attest, Power is not bound by reason or law, or by need or desire. It is unbound.

It is the only *Power* required to image *sensations* in accord with *ideas*—to *manifest* this whole '*Play of Life*'.

In religious terms, this *Power* is called *Omnipotence.*

Power is not controlled by Consciousness.
It is a facet of Consciousness.

Power cannot be *Sensed* (for it is not a *sensation*).
Nor can it be *Known* (for it is not an *idea*).

From an objective viewpoint, *Power* cannot be directly discerned.
It can only be inferred from its effects—the *orderly arrangement* of *sensations* and *ideas* within *Awareness.*

As Consciousness spontaneously determines,
Power acts—instantly manifesting experience.

This is evident in something as simple as 'tying your shoelaces'.

***

Consciousness does not *Know* 'how' *sensations* and *ideas* arise together to manifest the *appearance of hands* within the *visual field*, how they perform their *apparent intricate dance*, or how the *attendant feelings arise.*

They simply do—as soon as the intention appears.

What is true of tying laces is true of all manifestation.

***

What *Power Is*—and 'how' it does what it does—remains a *conceptual* mystery to *Consciousness*, as any *Known*-'how' would be an *idea*, and…

Power is not an *idea.*

It is *Real.*

Plainly, 'how' is no mystery in *practice.*

Plainly, Power instantly manifests the appropriate set of *sensations* to accord with any and every possible set of *ideas* (in and to *Awareness)* to create this Play of Life—from every possible perspective… each in turn.

Consciousness and Illusion

10. Splitting *Non-dual Consciousness* in Two

Image courtesy to ChatGPT

As outlined in the sections above, the primary *facets* of *Consciousness* are:

- ***Sensing-sensations*** (*colours, odours, flavours, feelings, sounds*)
- ***Knowing-all-ideas*** (of *number, form, and meaning ever-unchanged*)
- ***Power*** to manifest *sensory images* in accord with *ideas* to create this 'Play of Life'… in and to its own *Awareness.*

Though *Non-dual*, Consciousness has two *aspects* that can be separately discerned: the *Changeless Reality* and the *ever-changing appearances.*

Take *Seeing-colours.*
Though inseparable, you can distinguish between:

- the *unchanging, clear Seeing,* and
- the *ever-changing, opaque colours.*

This becomes evident with a simple experiment.

Look through a tube wide enough for both eyes (or cup your thumbs against your nose and hands around your eyes).

Notice that by shifting attention away from the *coloured image* at the 'far end'…

At the 'near end', there is *nothing at all* to be *Seen.*

This 'near end' is the Seeing, which itself cannot be Seen.

Of course, in *Seeing*, there are no actual 'near' or 'far' ends. These words are simply used to direct *attention* to the *Seeing* itself.

The *colours* forming the inside of the *apparent* tube—or of your hands—are no closer to the *Seeing* than the *coloured scene* at the *apparent* 'far end'.

The entire tube, hands, and 'distant' image are all *appearances* within the one unbroken *visual field.*

This *field* does not extend in space.

There is no 'near or far', 'up or down', 'left or right'—except in *appearance.*

All space *appears in and to* this *Clear unchanging Seeing.*

***

While it's possible to distinguish the *Seeing* from the *seen*, as earlier observed: *colours cannot appear in the absence of the capacity to See.*

Seeing-colours is really an Inseparable Whole.
It is Non-dual.

What is true of *Seeing-colours* is equally true of: *Smelling-odours*, *Tasting-flavours*, *Hearing-sounds*, *Feeling-feelings*, and *Knowing-ideas*.

There is an *Unchanging,* and an *ever-changing,* aspect to them all.

It is the *capacity to Sense* that never changes, while the *sensations* constantly *appear* and *disappear* within *Awareness*.

Though named and separately noticed, these facets cannot be split off from Non-dual Consciousness, any more than you can extract facets from a diamond.

This *Reality* cannot be grasped by thought.

In *simply being Aware*,
it is already *self-evident*…

to *Consciousness* ☺

11. The *Appearance* of *Change* and *Time*

Image courtesy of ChatGPT

As just observed, this *clear Seeing* is *changeless*—for it has no features that could ever change.

Only the *colours appear* to change.

And *without change*, there can be no time.

***

The capacity to *Sense-sensations* never changes.
Nor does the *Knowing-all-ideas.*

It is precisely because Consciousness is always NowHere—Timelessly Present and Unchanging—that the appearance of change can be perceived, as sensations rise and fall.

Try this:
Hold up your hand and move it slowly across the *visual field.* Before the hand *appears*, while it is *visible*, and after it is gone… *no time has passed* in the *Seeing* itself.

THE DOOR TO NOWHERE

Time is not *Real*.

It is only a *concept*, measured by reference to *apparent* change *within* the *illusion*.

It has no *existence* beyond the *play of appearances*.

***

This contention will be further explored in Part Two: When Science Meets Awareness and in Part Three: Introspection.

12. *Duality* and *The Veil of Ignorance*

Image courtesy of ChatGPT

When we looked at Knowing ideas earlier, while it was plain that some *ideas* are *Known*…

It was just as clear: 'there is no direct *Awareness* of *Knowing all ideas*'.

But Notice:
There is no *Awareness* of *Knowing any idea*—until *Power* associates a specific set of *ideas* with some *sensory image*.

The image may *appear* as the apparent thing itself, a picture of the thing, a hallucination, dream, an imagined object, a thought…

Or a word referencing the thing:

Teapot.

From the viewpoint of the reader (and the writer), this word *appeared* suddenly—out of *NowHere*.

Not only does the word *appear*—its *meaning* is immediately understood.

Yet, before the word *appeared*, there was no *Awareness* of *Knowing*—or *not-Knowing*—the *idea* of *teapot.*

What is true of *teapot* is true of all *ideas*:

Until a sensory image (including 'subtle' sensory images experienced as 'thoughts' and 'imaginings') is formed by Power to represent the idea within and to Awareness…

It is as if the idea is not.

***

Apparent Duality is founded upon the ten 'Ignorances' of *Awareness,* explored in the prior sections.

These are aspects of Itself that *Awareness* is not usually *Aware* of. Perhaps, 'you' (as *Awareness*) now recognize them:

- Ignorance of any *sensation* until it appears in *Awareness*
- Ignorance of *Knowing-all-ideas, ever-unchanged*
- Ignorance of *any idea* until it is *paired* with a *sensory image*
- Ignorance of *how Power works* to pair *ideas* and *images*
- Ignorance that all *appearances* are made of *sensations* and *ideas.*
- Ignorance that *appearances* are *immaterial.*
- Ignorance that *appearances* are inseparable from *Awareness.*
- Ignorance that *Awareness is immaterial*
- Ignorance that *Awareness alone Is.*
- Ignorance that *Awareness* is not divided among 'beings'.

Essentially, what is not in Awareness… is as if it never exists.

From this *ignorance*, different *sensations* and *ideas* together *appear* in and to *Awareness* to be: 'me' (my body, thoughts, and feelings) and 'the world' (everything else)… giving birth to *apparent* 'duality'.

13. *One Life*, Many *Appearances*

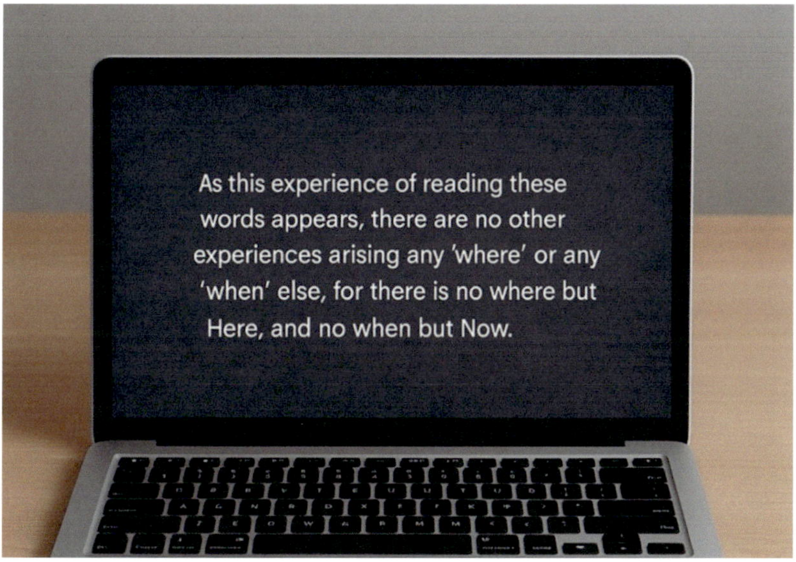

Image courtesy of ChatGPT

Where *One* is…*One* is wholly *Present.*

Consciousness is never 'split into bits'—some for 'you', some for 'me'.

Every apparent being is an illusion—no different to a dream. Dream beings cannot 'contain' consciousness. They *appear* in Consciousness.

It follows that:

As this experience—of reading these words—appears,
there are no other experiences happening
any when or where else.

For there is no 'when' but Now,
and no 'where' but Here.

What is true of this *experience* is true of all *experience—as each arises*, there are *no others appearing.*

Between one *experience* and the next, no time really passes—for *nothing ever really changes.*

***

Every experience appears in and to Consciousness Alone.

In religious terms, this *presence* in which and to which all things *appear* is called *Omnipresence.*

***

Yet, this is not the same as 'solipsism'—the *belief* (*idea*) that 'I' am the only 'mind' or 'self' that exists…

14. *Apparent Life* is as *Real* as it Gets

Image courtesy of ChatGPT

Without *sensations* and *ideas* manifesting a present memory of any 'other life' in *Awareness*, each *apparent life* seems to be 'the only life I've ever lived'.

It is self-evident that…

I (the reader) have no memory (no *idea*) of being I (the writer).

Nor do I (the reader) have any memory or current *Awareness* of the 'thoughts and feelings' of any 'others'. As only 'my thoughts and feelings' are *present NowHere.*

This is true whichever 'me' I appear to be *NowHere.*

***

This is not solipsism for the simple reason that while *apparent* 'bodily' life is an *illusion*, the *experience* of each and every life cannot be more *Real.*

THE DOOR TO NOWHERE

Just as *you* feel yourself to be a separate person. So it seems from the viewpoint of *every* being. Each having their own apparent body and mind. Their own identity and life history, separate from all others.

Each being is as real to themselves…
as you are to yourself.

The people you know and love (or hate) are not simply 'lumps of dead matter' walking around like zombies—nor mere 'dream phantoms'.

The *One* that seems *Now* to look out through the eyes of the *apparent reader* of these words...

Is the same that seems to look out through the eyes of the *apparent writer* —as the words are written. And…

Is the same that seems to look through the eyes of every *apparent being— each in turn.*

***

Thus, *experienced life* is not solipsistic, for:

What is seemingly done to another being
is truly Felt from the viewpoint of that being.

Hence… the Golden Rule. ☺

15. *The Illusory Universe* is *Perfect*

Image courtesy of ChatGPT

Only *sensations are Sensed*, and only *ideas are Known.*

That they *appear* to be any 'thing' at all—is not a mistake, nor a delusion.

It is not a trap.

It is the *Great Illusion—it* is how *Reality appears* to *Itself.*

***

From the human viewpoint, the *apparent world* often seems full of strife, suffering, and terror—something to be fixed or avoided.

Yet… all is perfect as it Is.

Reality is in no need of improvement.

At first glance, this may seem cold and uncaring—especially considering the immense suffering in the world, and the *pain* felt in every person's life.

But reflect more deeply:

- You cannot have valour without cowardice.
- No heroism without villainy.
- No light without shadow.

Every duality defines its opposite. As it is said:

The height of the mountain…
is measured by the depth of the valley

The sharper the sorrow, the more radiant the joy—once it is transcended.

With suffering comes compassion.
With difficulty, courage.
With resistance, patience…

Virtues and vices are not mere abstractions.
They are forged in Experience.

This *apparently imperfect universe* is 'perfectly imperfect'.

Not just because it gives form to all dualities—manifesting this 'Play of Life'—but also because, ultimately…

No one is ever truly harmed.

The instant pain and suffering are gone—
it is as if they never were.

Right *Now*, reflect on some pain or sorrow once deeply felt, and forgotten—until *Now,* as the shadow of it is recalled to memory.

Is it not so, that—until this moment of recollection—it was as if it never happened? Every pain shall pass.

Though that hardly seems like consolation as it is *experienced!*

Consciousness and Religion

16. Religious *Revelation*

Image courtesy of ChatGPT

As alluded to in the Preface: *God, Allah, Suchness, Brahman, the Tao, Consciousness* — are all poetic names for this *One Reality.*

***

In **The Awakening of Faith, a Mahayana Buddhist treatise**, it is written: *"The True Reality is originally only One. Its true nature is that of a pure mind—eternally joyful, the true being of all things, pure, quiet, unchanged, free, divine, unending, unchanging, and unspeakable."*

In the **Qur'an (Surahs VII:200 and XLII:11),** it is revealed: *"He is Hearer, Knower… the Hearer, the Seer."*

***

The Hindu Brihadaranyaka Upanishad (III.vii.23) declares: *"He is the unseen Seer, the unheard Hearer, the unthought Thinker. Other than He there is no hearer, no thinker, no comprehender."*

***

In Christian and Jewish tradition, God is described as:

***The One Invisible, Eternal, Unchanging, Living Spirit*:**

- **Romans 3:30** — *"God is One."*
- **Revelation 1:18** — *"I am the Living One!"*
- **John 4:24** — *"God is Spirit."*
- **1 Timothy 6:16** — *"He alone is immortal."*
- **Colossians 1:15–17** — *"… the Invisible God."*
- **1 John 4:12** — *"No one has ever seen God."*
- **1 Timothy 6:15** — *"No one can ever see Him."*

***In whom, to whom, and for whom the world is made manifest*:**

- **Romans 11:36** — *"All things were created by Him, and exist through Him and for Him."*
- **Isaiah 48:13** — *"When I summon the earth and sky, they come at once and present themselves."*
- **Lamentations 3:37** — *"The will of the Lord is always carried out. Good and evil alike take place at His command."*
- **Daniel 4:35** — *"He looks on the people of the earth as nothing."*
- **Psalm 39:5–6** — *"Indeed, every living man is no more than a puff of wind… no more than a shadow. All he does is for nothing."*

***

Part Four: A Religious Perspective explores in depth what the source texts of all the great religions reveal about *Reality*. And—contrary to general belief—what emerges is a broad and wonderful concurrence.

Consciousness is *Reality*

17. Consciousness Alone Is

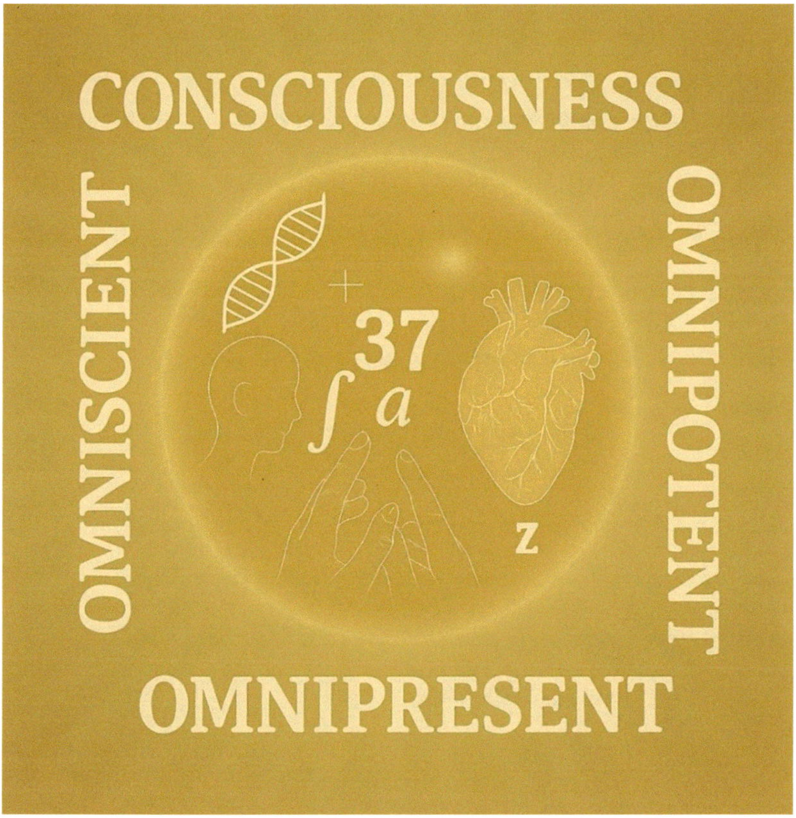

Image courtesy of ChatGPT

And so, we come full circle.

Consciousness is This: Invisible, Intangible, Inaudible, Inconceivable, Timeless, Changeless, Spaceless, Non-dual Reality…

- ***Sensing-sensations*** — *colours, flavours, odours, feelings, and sounds*
- ***Knowing-all-ideas*** — *number, form, and meaning, ever-unchanged*

With ***Power*** to create—moment by moment—the *appearance* of a solid, 3D world within and to its own *Awareness…*

THE DOOR TO NOWHERE

A world full of astonishing things and beings, across all scales—from the subatomic to the cosmic—including dreams, hallucinations, and stories of every kind.

***

Each *experience appears* from the 'first-person perspective' of an *apparent being* in the *apparent world*, as if looking out its eyes…

Experiencing its body, thoughts, feelings, and surroundings as:

"my body"… "my thoughts"… "my feelings"… "my world"…

As in a dream.

Where every interaction is replayed from each perspective, in turn…

Giving Life to all beings.

***

Alone *Knowing all ideas ever-unchanged*, it is called **Omniscient**.

Alone being *the Power that manifests all,* it is called **Omnipotent**.

Alone *ever-present in every experience*, it is called **Omnipresent**.

And—as incomprehensible as it may seem…

18. *This Is Thou*

So long as no *idea* is formed about the words: 'This', 'is', or 'Thou'.

Being *NowHere* is *recognition* enough…
(without 'thought' trying to define 'who' is *NowHere*).

Image courtesy of ChatGPT

The rest of this book is a direct invitation to observe—moment by moment:

the Sensory Images, the Ideas, and the Power—that together—manifest this ever-changing, illusory universe…

As One, within and to this Unchanging, Timeless, Non-dual Consciousness.

Only you can test what is being pointed to, as we look, together, at each facet in turn.

But first…

The Story-Teller

You may think that this is my story. Yet, I'm merely the *apparent* scribe. It is you alone who reads *meaning* into these words—as you read them.

It is you who are telling your own story. So…

What are You Thinking?

Perhaps thought has paused, if only briefly, as *Reality reveals Itself—to Itself*

Or perhaps, if you're reading with the mind of Richard Dawkins, Neil deGrasse Tyson, or Sam Harris, you remain sceptical. That's OK.

This book doesn't ask you to abandon science. Only to follow its own findings to where they inevitably lead.

To that end, we'll soon explore how science accounts for *perception* (both *sensation* and *meaning*) in much more depth, so you can test the *theories* for yourself.

***

We'll also revisit the world's sacred texts. Not to debate dogma, but to hear what they whisper to the heart once *ideas* are let go.

For the spiritual mind—Jewish, Christian, Muslim, Hindu, Buddhist, or from any other tradition—this book may be confronting.

But it does not dismiss faith or reason.

Rather, it invites both to ripen… into direct *Seeing*.

If scripture is the seed, let *Life Itself* be the fruit.

***

THE DOOR TO NOWHERE

Whether through sacred texts, the scientific method, or silent introspection…

All paths share a common aim:
to reveal what Is.

Which is the aim of this book. It does not begin with answers, but with questions.

It weaves together:

- a personal story,
- a scientific inquiry into perception,
- a direct examination of experience, and
- a return to the wisdom of scripture.

All converge on the insight that the *apparent universe* is an *illusion.*

And all *appearances* arise within and to this *One Reality*…

NowHere Aware of these words.

***

Still, you may be wondering…

Why Am I Asking About Your Thoughts?

First, to pre-empt some objections.

***

But there's another subtler, far more important reason:

To invite you to become *Aware* of your own thoughts.

To *See* how they construct stories—about 'you', about 'the universe', about 'what is'.

To **notice** the quiet miracle: that while *thoughts change*, *Awareness does not.*

***

And…

To **directly observe**:
that without thought, there is no story to tell you:

- who or what you are,
- where you are,
- when you are,
- how or why, you are…
  There is simply…

You Are…

As *I am…*

Part One: My Story

Having now glimpsed the terrain we return to the beginning.

To the moment when the journey's end apparently arrived for the author… and then continued-on.

I've chosen to not credit this book to the apparent writer, as it is not about 'me'… but about 'you'.

Nonetheless, it may help to ground the story if I tell you a little about myself.

An Ordinary Seeker.

I am no guru. I've never gone to India or sat cross-legged (I can't do it! ☺)—determined to become enlightened.

I'm just an ordinary Aussie bloke—nothing mystical.

My life has been both fortunate and unremarkable.

I was born into a conventional family in one of the richest countries on Earth. I had a normal 1950's middle-class childhood, a private education, then work and travel, and a bit of civic service, with a loving family and longstanding friends. There have been ups and downs, including six months of excruciating pain until a surgeon fixed my lower back, but overall, it's been a life of relative ease—rich beyond the dreams of most people.

***

Yet, for as long as I can remember, a quiet but persistent question haunted the background of my mind: Who am I?

It wasn't just philosophical musing, or a search for psychological solace.

It was an existential demand…

THE DOOR TO NOWHERE

An undercurrent moving beneath the surface of daily life.

What is this all about? Who, or what, is experiencing it…

And more deeply:

- What is the source of my consciousness?
- What is the purpose of being?
- Is there life after death?

And on a higher level:

- Is there a God?
- Why does pain exist and evil abound if this world is the creation of a good and loving God?
- How can there be so many religions and philosophies, yet only one Omnipotent God?
- How can my own free-will be reconciled with such a God?
- If the world is made up of atoms and molecules as science demonstrates, where is my spirit—the essential 'I', that *I feel* myself to be?

***

Some people search for answers in the world around them. Others turn inward. Science offers one view. Religion another. And in between are endless New Age teachings, each claiming to know 'The Way'. But where is *Truth* to be found?

I spent over a decade searching for it through philosophy, science, and religious texts. I read widely. I reflected deeply. But no answer ever came in the form of knowledge.

Instead, something altogether unexpected happened: the "me" that was searching simply vanished.

Epiphany

The End of 'Me'

One autumn day, over 40 years ago, as I raked leaves in the backyard, a thought arose:

"If you are determined to keep on this path, you will die."

And then, in an instant, the person I thought I was, disappeared.

Still standing there, rake in hand, the world I had known went blank. The familiar reference points of identity: my name, my history, my sense of self, all my surroundings—dissolved into nothingness.

There was no 'me' left at all.

When perception resumed, what returned was not someone looking out upon a world, but simply…

Pristine Awareness

Seeing—and Hearing and Tasting and Smelling and Feeling and Knowing.

There was no person in a body—only fleeting, ever-changing *sensory images* that appeared to form: *bits of my body*, *my thoughts*, *my feelings*, *my surroundings*, as specific *ideas* were associated with *images.*

The backyard was just as it appeared before—my body still standing, rake in hand. Yet everything was different.

The 'self' I had always assumed to be the experiencer had gone, leaving only *experiencing.*

That 'self' was seen for what it is—not the *subject* of experience, but just another *apparent object*, appearing within and to this…

Boundless, ever-present, unchanging, non-dual Awareness
that is NowHere.

The World of Illusion

The world, once so solid and dependable, was revealed to be as insubstantial as a laser light show projected onto smoke.

And yet—nothing was lost.

Being Truly Alive

Life itself has simply become more immediate. More intimate. More precious.

Not because 'I' have gained something, but because—with the illusion of ownership dissolved—experience is no longer 'mine.'

Going With the Flow

The old sense of urgency—of something missing—has gone. There is no longer a 'someone' trying to fix or manage 'my' life.

Life simply *Is*— unfolding effortlessly.

- Where once I searched for meaning, now there is wonder.
- Where once I struggled to control, now there is allowing.

Engaging in the World

What hasn't changed is the sense of 'me' as body and mind. It still appears, just as it always did. Still loving and engaging with family, and friends… and writing books ☺

And even continuing to work with a great group of 'kids' (in their 20's and 30's) who are trying to get a basic income implemented in Australia.

Yet, there is no longer any ownership of 'outcomes'.

Thoughts, emotions, and feelings have continued to arise, including—many years after the 'dissolution' in the backyard—months of intense pain before my back surgery, along with deep anguish when my wife nearly died several years ago.

But the difference?

The past does not cling to me. Once it is gone, it is gone.

The whole point of Life is to Live… NowHere!

A Lifelong Journey

After that shift in perception, when it became clear there is only *Awareness*—a quiet compulsion remained: to explore whether this same *Reality* had been revealed in the stories of others.

Reading Ancient Scripture

I turned to the scriptures—reading the Bible cover to cover three times, the Quran, the Bhagavad Gita, the Upanishads, and Buddhist sutras. I devoured philosophical texts and the words of sages and mystics…

I wasn't drawn to the stories themselves, but to the insights they offered into *Reality*.

Everywhere, the same truth echoed. But always, it seemed to come from some ancient source.

Exploring Modern 'Sages'

Until I encountered the words of Krishnamurti, Nisargadatta, Alan Watts, Douglas Harding[7]—and many others—all pointing directly to what had already become plain.

The deeper I looked, the clearer it became: this realisation was not unique.

It has always been NowHere—whispered across centuries, within every culture from every corner of the world.

And then the internet exploded...

I discovered Tony Parsons, Rupert Spira, Eckhart Tolle—even Sailor Bob and his wife Kat Adamson, along with Joseph Blake here in Australia. Many others across the world too.

---

[7] Douglas E. Harding, *On Having No Head: Zen and the Rediscovery of the Obvious* (London: The Buddhist Publishing Group, 1961)

All spreading the same message in YouTube videos, books, blogs, posts, podcasts, and Satsangs.

There are even Facebook groups—'No-Facebook'[8] and Miranda Thingy's 'Nonduality Dreamlife'—with thousands of members, along with a website: headless.org[9]…

All proclaiming the *Non-dual* nature of *Reality*.

Exploring Modern Science

At the same time, I burrowed into scientific papers about the *theorised* forms and laws governing the *material* universe, along with the findings of neuroscience…

Which seemed only to confirm—the illusory nature of the perceived universe.

Simply Observing

In parallel, I observed.

Seeing, Hearing, Tasting, Smelling, Feeling, and Knowing…

Perceiving the illusory nature of all appearances—beyond doubt.

The traces I followed are traced again in Part Three: Introspection.

Re-telling the Tale

Looking back, it seems that all the reading and watching arose out of simple joy—of seeing the same Truth expressed again and again—in different words by different voices.

[8] Promoting Douglas Harding's insights
[9] Also reflecting Harding's viewpoint.

Eventually, the joy turned to writing—articles on Medium, posts on Quora, and now… this book.

Telling the same story, but from a unique perspective.

While *Awareness* Remains Ever as It *Is*

Nothing I've read, heard, or observed has added to—or diminished—this *ever-present Awareness.*

It is always already complete. *Self-evident.*

For there never was a seeker, nor anything lost that needed to be found.

Which raises the question…

What of Birth and Death?

Living the Dream: Birth and Death

My life-story started, of course, with my birth—in 1949.

Mum once told me—years later—that when I was born, she felt as though she'd given birth to Christ. That always gave us both a good laugh.

Though, as the later epiphany, and my reading of scripture, revealed… in a way, it's true of everyone.

A few years ago, Mum died at the age of 97.

In her later years, she would often reflect on the hope of seeing her own mum and dad again in 'the afterlife,' along with her husband (my dad), and other friends and relatives long since passed.

I would tease her…

How old do you think they'll all be? Would everyone appear at the age they had died—even babies and children, and the elderly?

Or would everyone return in their prime, even the kids?

But that raised other problems—what about the 'ugly' and the 'beautiful'?

Would everyone appear just as they did in life—for all eternity?

Then there were second and third marriages, stepchildren, extended families… all the way back to the dawn of humanity. Not to mention all those yet to be born.

It started to sound like quite the melee—everyone trying to catch up with (or avoid) everyone else!

Still, despite the teasing, Mum never lost her hope—even on her deathbed, as we held hands and talked.

***

Until, quite out of the blue, her final words to me were…

"Strange how it is always the same One who seems to die."

Ironic, given her longing for an afterlife filled with others.

Yet perfectly in tune with the revelation that…

There is indeed only One Living Reality…
That never really dies.

Good News indeed. ☺

As promised in the Preface—this journey doesn't ask you to abandon science or your faith. It may, however… transfigure them both.

Part Two: When Science Meets *Awareness*

This Part explores in detail the scientific theory of *perception* touched upon in The Road Map.

You can skip ahead to Concepts of Space and Time on page 73, if you already recognise the **Paradox** that the theory poses.

Emergence of *Perception* from *Matter*

According to science, without *Consciousness*—without *Awareness*—there are *no colours, no flavours, no feelings, no odours, no sounds, and no ideas (of number, form, and meaning).*

These do not exist in the so-called *physical* universe.

But don't take it from me. Let's explore it together.

To simplify the investigation, let's first consider just one facet of *Consciousness*: *Hearing-sounds.*

What follows is a *theoretical description* of how *sound* is said to *propagate* in the *material* model and how it is said to '*emerge*[10]' *from physical processes*—so we can better contrast it with *direct experience.*

There is a great deal of detail, for a reason. It is to show that there is no 'simple connection' between the *theorised* '*physical universe*', and the '*perceived world*' that is *directly experienced.*

[10] *Emergence* is the *idea* that a complex system (like a brain) can give rise to new properties (*Seeing-colours, Feeling-feelings and Knowing-ideas*) that are not evident in its individual parts. But this does not *explain* how *immaterial experience* arises from *material activity*. It just claims that it does.

Hearing-sounds

Outside the *Material Body* (in the *Material World*)

Let's say a drum is struck. This causes the skin of the drum to vibrate. That's a mechanical movement—but it's not *sound.* The vibration creates a pressure wave in the surrounding air—molecules bumping and shifting in a pattern. This wave isn't *sound* either—just molecules in motion, no different in essence from the vibrating molecules in the drum. The air wave may hit a microphone, making its diaphragm vibrate. Still not *sound.* This moves a coil. That movement is not *sound.* The coil moves through a magnetic field, generating an electrical signal. That signal is not *sound*

The electricity flows into a transmitter antenna, making electrons oscillate. Still not *sound.* This generates radio waves—electromagnetic ripples that travel through space. Again: not *sound.* In fact, these waves are just like light, only with a longer wavelength! When the radio waves reach a receiver antenna, they make electrons move—creating another electrical current. This current is *soundless*, too.

The current may power a speaker, causing a coil to move. Once again, mere movement is not *sound.* The coil moves the speaker cone—still just mechanical motion that is itself *soundless.* The cone pushes on air molecules, creating another pressure wave. Once again: not *sound.*

At every step, we see different forms of matter and energy in motion. One of these forms—radio waves—can even travel through airless space at the speed of light.

Nowhere in this journey is a single sound-Heard.

Inside the *Material Body*

Next, the air pressure wave reaches an eardrum, making it vibrate. This vibration is not *sound.* The vibration is passed to three tiny bones (the malleus, incus, and stapes), which amplify the movement and… pass it to the oval window (fenestra ovalis), a thin membrane at the junction of the middle and inner ear. Still—no—*sound.*

The vibrations ripple through fluid in the cochlea (a snail-shaped structure in the inner ear). The ripples too are not *sound.* A wave forms along the basilar membrane inside the cochlea. Nor is this wave making any *sound.*

Hair cells riding the wave cause microscopic projections (stereocilia) to bend against an overlying structure. This bending is mechanical. It's still not *sound.* As the stereocilia bend, they open tiny channels. Chemicals rush in. The rushing chemicals make no *sound.* This chemical movement generates an electrical signal. Nor is this 'signal' a *sound.* That signal is transmitted via the auditory nerve as an electrochemical impulse to the brain. Still, not *sound.*

At no point—from drum to inside the brain—is any sound-Heard.

Only physical processes are theorised to exist in the physical universe.

Inside the *Material Auditory Cortex*

Finally, the electrochemical impulse reaches the auditory cortex, and in *complete silence deep inside the head…*

Somehow—no one knows how— this *theoretical* signal is *theorised* to be transmuted into…

- An *actual sound*
- that is *Heard.*

This cannot be tested objectively, as the *sound* theorised to arise inside the head cannot be *Heard* by any third-party observer. Nor can it be detected by any scan or instrument, which only detects 'electro-chemical impulses' or other *physical correlates* of some sort.

Sound can only be Heard subjectively.

Which raises two questions:

- Where is the *sound*?
- Who is the *Hearer*?

According to the *theory*, *Hearing-sounds* arises only at the end of the long *material process* described above—as an *inseparable* (*non-dual*), *immaterial 'emergent' phenomenon* within the *material auditory cortex*.

Seeing-colours

As foreshadowed in the Road Map, what is true of *Hearing-sounds* is true of *Seeing-colours*.

Light, as electromagnetic energy enters the eye and triggers an electro-chemical response in the cones and rods of the retina—*if* it falls within the narrow "visible" spectrum.

The electro-chemical energy is then *theorised* to travel along the *material* optic nerves to the *material* visual cortex at the *back* of the brain.

This *physical* flow of electrochemical energy is *not* what we *See* as *immaterial colour*.

Instead, inside the head, in complete darkness, the electro-chemical pulse is *theorised* to be transmuted (once again, no one knows how) into actual 'patches of *immaterial colour*', and the actual '*immaterial Seeing experience*'.

On this theory, there is no 'little person' inside the head looking out through the eyes and into the *physical universe*.

The optic nerves are opaque, and all the signals are theorised to flow one way—outside to inside.

It means that (on this theory) when you look…

All that is apparent is a momentary immaterial coloured image—inseparable from the momentary immaterial Seeing, in which, and to which, the image appears.

This conclusion may seem counter to all *experience.*

Yet it is *observable* whenever you look at the night sky.

***

Light is said to travel through space at an absolute speed of around 300,000 kilometres per second.

On a human scale, this is virtually instantaneous, yet from a scientific perspective, there must always be a time delay, no matter how short, between light leaving a *physical object* and *perception of the object.*

Why is the time delay so important?

Because it helps to demonstrate that:

'What' is 'Seen', cannot be 'physical objects' composed of atoms and molecules.

This is more easily recognised if we follow light from a supernova (a star that has exploded).

In this case, the light travels through the cosmos during a period when the star has already ceased to exist—but before any *image* is *Seen.* Showing that the *perception of the image* does not depend upon the existence of any *physical object* at the time the *image* is apprehended.

On the *material theory*, as the observer and source move closer together…

The only effect is that the time delay shrinks.

From this it follows that being close to a '*physical object*' cannot change the nature of what is *Seen*, namely…

A 'momentary immaterial coloured image' inseparable from the momentary Seeing experience.

Not *the 'physical thing itself'*.

***

Because this view is so contrary to the normal view, that what we *See* is the 'physical world', it may be worth considering another analogy.

The Eye as a Camera

We can imagine a 'light ray'—which is a theoretical construct—reflecting off an object in the theorised physical universe. It travels through space until it passes through the lens of the camera, interacts with a photoelectric device, is converted to an electric current which is transmitted along a wire to the central processor, where it triggers a process that illuminates a pixel on a display screen. Combined with many such rays, the result is a *coloured image* on the screen.

No one mistakes the *image* on a video screen for the *physical object*, nor expects to peer through the screen into the 'real world' behind it.

In scientific terms, what is true of the video camera is true of any visual cortex because the sequence of events is similar.

Based on this analogy, the *original source* and the *image* are not the same thing.

In scientific terms:

Vision reveals only coloured images—not the physical objects themselves.

Finally, your dreams also attest to the fact that you can *See* in the absence of any light.

In every case, *whether in a dream or in daily life:*

The colours themselves are the illumination,
inseparable from the Seeing.

Sensing-sensations

What is true of *Hearing-sounds* and *Seeing-colours* is equally true of *Feeling-feelings*, *Tasting-flavours*, and *Smelling-odours.*

According to physics and neuroscience, all *sensory images* are said to arise—*not as properties of the external world*—but only at the final stage of extended *material processes* that culminate inside the *material brain.*

Yet, no sensation can be found anywhere
within the brain itself.

Objectively, *sensations* are not located in any substance, nor in the nerve signals, nor in the synapses or cells.

They are only experienced subjectively.

For example: *feelings* don't exist in the *material universe.* You can't grab one—you can only *Feel* it.

And that requires *Awareness.*

Knowing-ideas

But what about *ideas—of number, form, and meaning?*

- There are no *immaterial ideas* in the *physical universe.*
- There are only *theorised material elements*: fields, atoms, molecules, and so on.

So, *what is* an *idea*, and where do they come from?

While there are various theories (touched on in Appendix 1), none show how *meaning* could arise from *meaninglessness.*

This question is best approached through direct observation.

It was briefly canvassed in *The Road Map* and will be explored more deeply in Part Three: Introspection.

Qualia as Subjective Experience

The term 'qualia' is used in science and philosophy to refer to the *subjective nature* of *conscious experience.*

Qualia consist of:

- ***Immaterial Sensations***—that which is *Sensed*, and
- ***Immaterial Ideas*** (or 'concepts')—that which is *Known*

Somehow (and no one knows how), these *immaterial sensations* and *ideas* appear *together*—within *immaterial Awareness*—to give rise to…

The *integrated immaterial experience* of: 'things, properties, processes, events, and relations'… that appear as 'me' and 'the world'.

Such as:

"Awareness of being a specific person reading this book…
in whatever place it now appears you are."

The Physical and the Psychical

We can now distinguish between:

- ***Theorised* physical inputs**—such as sound waves, flavour molecules, light waves, and electrochemical signals in the body— which can never be directly observed, and…

- ***Experienced* psychical qualia**— *immaterial sensations and ideas*, all of which *appear inseparably* within and to *immaterial Awareness* as the '*objects of experience*'.

The Mystery of *Signals* to *Qualia* to *Action*

There are two main *materialist hypotheses* that attempt to explain action:

1. ***Immaterial Thoughts* Cause *Physical Actions***
   This is the everyday view: *thoughts and intentions* drive bodily movements in response to what is *Perceived.*
   But this raises a deep mystery:

   How could something immaterial—a thought—cause something material to move, like a muscle to contract?

   ---

2. ***Physical Processes* Cause *Immaterial Thoughts***
   This is the dominant scientific view: *immaterial thoughts and intentions* are said to be mere by-products—*epiphenomena*—of *material* neural activity, with no causal power of their own.

   In some experiments, neural activity has been measured *before* a person becomes aware of deciding to act…
   As though the body acts on its own—like a zombie—and *Awareness* merely *observes*, after the fact.

   But this just flips the mystery.

   How could something material (like a firing neuron) give rise to something immaterial (like a thought or a feeling, or other sensation)?

In both cases, the question remains unresolved.

Storing and Recalling Memories

This raises the same mystery we've already seen in the link between *Awareness* and action.

How do *immaterial experiences* get *transmuted* into patterns of *physical states* that store a vast, interconnected web of memories, values, and beliefs—so that the stored *physical states* reappear in *Awareness* as coherent *immaterial memories or beliefs*?

This same problem extends into communication.

Mirroring *Meaning* in Communication

For communication to work at all, the *mapping* between *physical signals* and *immaterial meanings* must be *consistent across individuals.*

Whether *qualia* are *emergent by-products* or are *fundamental determinants*, communication depends on a reliable loop from *idea* to *action*, and back again:

- On the *first theory*, the speaker's *immaterial thoughts* are translated into precise *material muscular movements*—modulating breath, larynx, lips, and tongue to generate very precise pressure waves, as a person decides to speak.

- On the *second theory*, the *idea* is merely an *emergent phenomenon*, arising from the *physical processes* that trigger speech.

Either way, the pressure waves from the mouth of the speaker travel to the hearer's eardrum, where they are converted into electrochemical signals in the auditory cortex.

- The hearer's brain must consult stored *physical patterns*—representing memories, values, and beliefs.

- Then (somehow) reconstruct the **speaker's** *immaterial idea* in the **hearer's** *immaterial Awareness.*

No *physicalist scientific theory* explains either:

- The *self-evident appearance* of *immaterial sensations* and *ideas*,
- Or the full loop from *immaterial idea → physical body → physical environment → physical body →* back to *immaterial idea.*

Bridging Laws Between the Physical and Psychical

Physical laws *only apply* to *physical* entities—not to *immaterial sensations, ideas, or Awareness.*

Whether the *immaterial causes the material*, or the *material causes the immaterial,* for the *material view* to hold, there must be a *reliable mapping* between the two.

Some form of *psycho-physical laws* must have been embedded within the very fabric of matter–energy–space–time from the outset—*ready to emerge when the necessary structures (e.g., brains) are in place to give them expression.*

Without some 'bridging' laws, it would be impossible to:

- perceive a *consistent and coherent universe*,
- take *meaningful action*,
- recall *meaningful memories*, or
- engage in *meaningful communication.*

Without such 'laws', the only outcome could be random 'noise'.

Emergent Sensations, Ideas and Awareness

Even if the laws are posited and not proven, they can't act in isolation.

On the *material* view, along with the psycho-physical laws, the *immaterial* capacity to *Sense-sensations* and to *Know-ideas*, (along with *the immaterial sensations and ideas themselves)* must have been latent in the very fabric of the energy-matter-space-time complex at its birth.

Which means *immaterial Consciousness* itself must somehow have been dormant in the *proto-material universe.*

And not some 'embryonic' consciousness that gets aggregated into more sophisticated forms as matter aggregates, but fully realisable from the start.

You cannot break Consciousness into bits—as both sensations and all ideas are inseparable from the nature of Consciousness. It is an inseparable whole.

The Unbridgeable Ontological Divide

Which brings us to the fundamental philosophical puzzle:
How could two entirely different *categories—the material and the immaterial*—interact?

This isn't merely a problem of cause and effect—it's a problem of *category mismatch.*

There is simply no conceivable way to bridge this ontological divide.

But an even deeper difficulty lies within the *assumptions of materialism* itself.

The *Observer* is Never Inside the Observed

Even in physicalist terms, no entity sits inside your head observing the 'external' world.

Instead, the hypothesis asserts that *physical* flows alone—neural activity, blood chemistry, electrochemical patterns—somehow generate the *immaterial appearance* of:

- a body with a head,
- a thought,
- an emotion,
- a world, **and…**

The *Awareness* in which and to which these all *appear.*

In other words: inside the *assumed material brain* …

Observer and observed are Non-dual (Not-Two)

And with this…

We return to the **Paradox** at the heart of all *material perception.*

Paradox of Material Perception

As revealed in The Road Map, the perceived universe *appears* to be physical—and science treats it accordingly.

But *Awareness* never *perceives* 'matter' (as it supposedly exists)—only the *appearance of matter.*

The so-called *material universe* forever remains a *supposition*: an explanatory model *inferred* from *consistent patterns* within *Perceptual experience.* Never something that *Awareness* can step outside itself to verify.

Behaviour Does Not Require Awareness

Which raises another fundamental question.

If the world is entirely *material*—why *Consciousness* at all?

This question has no answer.

As, within the *physical world,* all *physical life-forms* can react perfectly to their *physical environment* by simply following the *physical laws*—without any need for *Awareness.*

What Are You *Thinking* Now?

Whatever you *believe* about who you are,
or what you take 'the world' to be…

Isn't it still the case that—you are simply *Aware*?

Let's begin from NowHere.
From this Ground of Awareness itself…

and follow how even science stands upon Awareness…

and See where that foundation leads.

Concepts of Space and Time

To place science in its proper spatial and temporal context, we must first recognise the different *concepts* of *space* and *time* that arise from the *material* view of *Reality*.

Remember: these are only *concepts*.
Not *Reality*.

Three *Concepts* of Space

With the understanding that the *apparent universe* arises as *immaterial sensations* and *ideas*…

We can now identify three distinct—and necessary—*concepts* of space required to account for science's treatment of *matter* and *Awareness*.

- ***Perceptual* Space** – what you experience.
- ***Physical* Space** – what you assume
- ***Ideal* Space** – what you know

Perceptual Space

This is the space directly *perceived* within and by *Awareness*—what you *See*.

It *appears* as a three-dimensional world, filled with 3D forms, but in *Truth*, it is a world of *appearances* only—*without substance or extension.*

It is the space that seems to stretch out before you:

- Small when looking into cupped hands.
- Vast when gazing at the night sky.

It encompasses the *illusory universe* explored in the Road Map.

After all, it's plain you can't fit an entire 3D world inside an *assumed* head!

In *Truth*, all *perceived* forms are *two-dimensional projections*[11] of *ideal* 3D forms—distorted by perspective and foreshortening.

The difference between *experienced Perceptual Space*, *assumed Physical Space*, and *Known Ideal Space* is easily *Seen* using the humble ruler.

In *Perceptual Space*, a one-metre ruler may *appear* taller than a mountain—clearly distinguishing *appearance* from *assumption* or *idea* (where the ruler is always a 'standard size' and much smaller than any mountain.)

Physical Space

This is the three-dimensional realm *assumed* to exist outside *Awareness.*

It is said to contain all *material* fields and forms—
But it is entirely *theoretical*—a *conceptual assumption.*

Even in *theory*, it contains:

- no *sensations*
- no *ideas*, and
- no *Awareness…*

Only hypothetical 'stuff' (mass and energy) in a *theorised* 3D Spacetime.

Most people—scientists included—believe they are observing Physical Space.

Yet, in following the *theory* to its logical end in the **Paradox**, it should now be clear that what is *observed* is always—and only—*Perceptual Space.*

[11] A 2D projection shows how a 3D object looks on a 2D surface, preserving depth cues through techniques like perspective, shading, and foreshortening, but without actual depth. It's why we can illustrate familiar forms on a 2D computer screen so that they appear to be 3D. This is especially evident with 3D Virtual Reality images.

Ideal Space

Ideal Space—like all *ideas*—is *inseparable from Knowing*, a *self-evident capacity* of *Awareness*[12].

It is the abstract *conceptual space* used to describe and measure what *appears* in *Perceptual Space.*

It is where mathematics and geometries exist—and is the context for all scientific models of the universe.

Here, *ideal* 3D forms are defined (e.g., the perfect cube), and *ideal* distances and velocities are measured.

In *Ideal* Space, the *ideal metre* ruler is used to measure the *ideal* mountain's height.

Summary of the Three 'Spaces'

- ***Perceptual* Space** – *what appears* in and to *Awareness—as directly experienced.*
- ***Physical* Space** – what *is theorised* to exist independently of *Awareness*—never observed, only *assumed.*
- ***Ideal* Space** – the *conceptual model—inseparable from Knowing—* that holds all *ideal 3D* forms and their relationships.
  It maps *Physical Space*, but without the *assumed* substance.

[12] To be explored in detail in Part Three: Introspection

Three *Concepts* of Time

Several theories of time will be discussed in Part Three: Introspection. For now…

We can mirror the structure of space to identify three concepts of time:

- ***Perceptual* Time** – what you experience
- ***Physical* Time** – what you assume
- ***Ideal* Time** – what you know

Perceptual Time

Perceptual Time is the *apparent* flow of time as it is directly *experienced*—what seems to happen now… and now… and now again. It is inseparable from change, memory, and anticipation.

It is *perceived* as a sequence of *appearances*:

- A leaf falls…
- A person ages…
- A thought arises.

Yet, each is simply a *present appearance*—an ever-changing pattern of *sensations* and *ideas* in *Awareness.*

Only the *present* is ever *experienced*. The past and future never *appear*.

The *illusion* of a timeline arises only in memory or imagination.

Time can only be *measured* by some *apparent process*:
a ticking clock, a rising sun—that *appears* in and to *Awareness*.

Physical Time

Physical Time is not a 'thing', but the *assumed* temporal dimension of the *assumed material* universe.

It is the time most people believe they live in—the time believed to age bodies, to separate one event from another, to differentiate past and future from the present.

It is also said to contain the instant 'present' (between the past and future) in which *experience* exists.

It is the time said to have begun with the Big Bang, that stretches onward to the Heat Death of the universe.

But like *Physical Space, Physical Time* is never *directly experienced.*

It is inferred by *theory… which is Known in idea*

Ideal Time

Ideal Time is the abstract, mathematical framework, used to describe and order events.

Here, equations describe motion, acceleration, decay. Timelines map histories and futures.

But *Ideal Time* has no flow. It does not pass.

It is a *timeless concept*—an *idea* in *Consciousness*—that gives *apparent order* and *continuity* to *experience.*

Just as *Ideal Space* governs *apparent* location…

Ideal Time governs the *appearance* of duration and sequence in and to *Awareness.*

It's the 'time' you *Know*—when you '*Know* the time'.

Summary of the Three 'Times'

- ***Perceptual* Time** – how change seems to arise now in *Awareness.*
- ***Physical* Time** – the *assumed* timeline of the *material* universe.
- ***Ideal* Time** – the conceptual scaffolding of the storyline unfolding in and to *Awareness—inseparable from the Knowing that Knows it.*

What is *directly experienced* is always—and only—the rise and fall of *immaterial sensations* coupled with *ideas* that give the *appearance* of passing *Perceptual Time, within and to Awareness.*

Three *Concepts* of Spacetime

- ***Perceptual* Spacetime** = *Perceptual Space* + *Perceptual Time*
  This is the field of lived *experience.*
- ***Physical* Spacetime** = *Physical Space* + *Physical Time*
  This is the *theoretical* domain *assumed* to exist outside *Awareness.*
- ***Ideal* Spacetime** = *Ideal Space* + *Ideal Time*
  This is the *conceptual matrix* that dictates how 'things' and 'events' relate to each other.

By now, it should be clear that everything that is perceived is *illusory.*

A *transient appearance*, within *Perceptual Spacetime*, and understood in the context of *Ideal Spacetime*—apparent in and to *Awareness*… like a dream.

With Spacetime clarified, we can now bring all 'things' into view.

The *Ideal* Objective Universe

Let's be clear—because everything hinges on this.

What you perceive is not the physical universe. It's a *sensory rendering* of some hidden *unKnowable Ground*—made *apparent* in and to *Awareness.*

This *Ground* is equivalent to the *supposed physical universe*—but without the *assumed substance.* More like *software than hardware.*

It is not 'out there'.

It is simply *Known*—in and by *Consciousness*—as an *idea.* Hence it is called '*The Ideal Objective Universe (or Ideal Universe).*

As no *idea* can exist apart from the capacity to *Know*, we can regard…

This 'Knowing' capacity is the 'hardware' that supports the Ideal Universe 'software'.

The nature of this 'hardware' and 'software' is not discernible.

But neither is the nature of the *assumed material universe.*

In all cases, the Ground is un-Knowable.

Unlike the *assumed physical universe*—which must be *inferred*—the *Ideal Universe* is *self-evident.*

You Know you Know. And what is *Known are ideas*—about what *appears in Perceptual Space.*

These *ideas* form part of the *Ideal Universe—or else they could not map it.*

Just like the *theorised 'physical universe'*, the *Ideal Universe* includes the *ideal* forms of all things—along with their *ideal* properties, and relationships.

It is these *ideal* forms and laws that science *discovers*—or rather, *uncovers.*

We'll explore this *Knowing-ideas* more deeply in Part Three: Introspection.

The next few sections look at how our *ideas* about the universe are mapped to our *observations*…

As we examine the *Role of Scientific Theory.*

The Scope and Limits of Scientific Theory

Paradoxically, recognising the **Paradox of Material Perception** does not diminish the role—or the discoveries—of science. At least, not as they relate to the *apparent immaterial universe.*

Scientific theories remain powerful within *Perceptual Space*—the domain of *lived experience.*

They simply no longer require the *assumption* of a *material universe* outside *Consciousness.*

***

Dreams show it is possible to *experience* a world that seems solid, fully illuminated, and governed by rules (of sorts)—even though it has no 'external reality'.

***

The 'dream world' and the 'waking world' share the same *immaterial quality*. They both *appear* in and to the same *immaterial Awareness.*

What differs are the *laws that govern the behaviour of apparent objects.*

In dreams, strange things may happen without rhyme or reason.

In waking life, the 'laws of nature' seem more consistent and predictable. It is these laws that science discovers through observation—in *Perceptual Space-Time*—and *models conceptually* with reference to *Ideal Space-Time.*

Letting go of the *material assumption* doesn't undermine the scientific method.

Mapping Theory to Observations

As outlined in *The Road Map*, the most rigorous scientific theories use mathematics to *model* the ***form*** of ***theoretical*** **objects**:

- from quantum fields and subatomic particles
- to atoms, molecules, proteins, and cells
- all the way up to stars, galaxies, clusters,
- and the cosmic background radiation.

Along with their:

- *Theoretical* **properties**—mass, charge, spin, etc.
- *Theoretical* **constants**—like the fine-structure constant.
- *Theoretical* **laws**—such as the conservation of momentum.

These describe the *theoretical* **behaviour** of all *theoretical* **forms**.

There is usually a long chain of mathematics linking *theory* to *observation*—including the mathematics *embedded* in the *apparent* devices used to measure, as well as in the algorithms that display the results.

Science does not require any material assumption to 'work'.

A theory is held to be valid when the ***theoretical behaviour*** **of the** ***theoretical forms*** reliably—though not necessarily perfectly—**maps onto** the ***observed behaviour*** **of** ***observed forms***: (observed in *Perceptual Space*, in and by *Awareness*):

- an *apparent* falling apple,
- an *apparent* graph on an *apparent* computer screen,
- or *apparent* tissue under an *apparent* microscope.

This is not to say that the laws discovered through the scientific method are the *actual laws* governing *appearances*. Only that they 'map' the laws in some respects. And are thus subject to change as new evidence *appears*.

While that is all… it is not trifling.

Revealing the *Ideal Universe*

It is—*in part*—through science that the vast depth, breadth, and interconnectedness of all 'things', on every scale throughout all space and time—within the *Ideal Universe*—is revealed in and to *Awareness.*

Science not only illuminates this intricately woven fabric of *Ideal* space-time-energy-matter but also enables the evolution of astounding technologies within the *apparent universe.*

These technologies, in turn, deepen understanding of the *Ideal Universe*—through physics, cosmology, geology, chemistry, and biology.

And through their expressive offspring: documentaries, animations, and yes—even YouTube videos!

***

'Doing maths and science' is how *Consciousness* reveals something of its *Omniscient* nature to *Itself.* Making the hidden apparent.

***

Nevertheless, as wondrous as it is, science has its limits…

The Limits of Scientific Theory

The following reflections highlight the inherent limitations of mainstream scientific *theories* about *Awareness.*

The purpose is not to debate the validity or utility of specific models—many of which are extremely useful within their domain—but to dispel the *assumption …*

That any model can reveal the nature of Reality *itself.*

It cannot.

Or that any model requires a material basis

It does not.

The Fundamental Limitation of Most Scientific Theories

By overlooking the **Paradox**, most scientists take it for granted that the *perceived* universe *is* the *material* universe.

From this *assumption*, they conclude that *Reality* consists of:

- Subatomic fields and particles,
- Quantum fluctuations,
- Emergent complexity, as well as
- Informational processes evolving in time.

But these are simply 'behavioural models'. They say nothing about the *Reality* that gives rise to *experience.*

***

We have used the computer analogy before, and will now do so again, as it is a powerful metaphor.

Doing science is like working out the rules of a computer game by watching how *images behave* on the screen.

From this study, you might learn to predict what happens 'on the screen', but you would never discover the nature of the images themselves, nor the source that generates them.

Similarly, science can uncover the laws governing the *form* and *behaviour* of *appearances*—because *appearances are observable.*

But that is all.

Just as you cannot get to the nature of the computer by observing images on the screen, science cannot reveal:

- what 'real life' *appearances are in 'essence'*, or
- the *nature* of the *Reality* that manifests and *perceives* them.

Which is to say, no 'scientist' can reach the heart of *Awareness*—simply because it is *Awareness* doing the looking!

Nor can any 'scientist' examine the *assumed material world, as Awareness* cannot step outside itself to see what is 'out there'.

Impossible to Acquire *Knowledge* Outside *Awareness*

Even *if* we *assume* the existence of an *external, material universe.* A deeper mystery remains:

How could 'dead' atoms inside a material brain ever come to 'know' anything about the world outside?

They cannot leave the body to go and look.

Shall we suppose, for example, that photons fleeing from a blazing building somehow "know" what the fire is? Or where it is?

Clearly not.

They are said to be blind, deaf, insentient particles—unaware of where they've come from, where they're going, or even that they are.

When such a photon strikes the retina of a passerby, what could it *knowingly* convey?

In *material theory*, the interaction between the photon and the atoms of the retina is entirely *physical—no sensation or meaning exists* at that point.

Just a cascade of electrochemical flows. With each atom in the body said to obey *fixed physical* laws.

Blindly.

Like a row of people passing along a push.
None knowing they were pushed.
None knowing they are pushing.
None even knowing they exist.

How could this *unknowing* chain of *physical* events ever produce:

- The *immaterial colours, tastes, smells, sounds, or feelings, and*
- The *immaterial meaning* of objects, events, and relationships?
- Along with *immaterial Awareness* of the *sensations* and *ideas* that together *appear* as a 'blazing building'

In short:

How does the ***unknowing*** *give rise to* ***Knowing****?*
How does the ***meaningless*** *produce* ***meaning****?*
How does the ***blind*** *become* ***Aware***

These questions remain unanswered—not for lack of trying—but because the very framework of *materialism* lacks the category required to address them.

As there are no *ideas* in the *physical universe*, no *qualia*, no *meaning*, no *Awareness*—there is no bridge, or mechanism, by which the merely *physical* could ever *become* the *psychical.*

Impossible to Perceive Change via a Changing Brain

The *direct experience* of *movement* reveals another deep flaw in the *theory* that *Awareness* arises from *changing brain activity.*

The flaw lies in the *conception of time.*

While there are several competing theories:

- Time flows one way, driven by entropy (*classical physics*)
- Past, present, future all coexist in a 'block universe' (*relativity*)
- Time is observer-dependent or emergent (*quantum theory*)

There is broad agreement that, *from the observer's standpoint*, the *present* is an *instant…*

An infinitesimal *now*, suspended between past and future.

***

Now consider, if:

1. The *present* is only a single *instant*, and
2. *Sensations* and *thoughts* are said to arise solely from *changing physical processes*, and
3. *Immaterial Awareness* is assumed to *emerge with those changes…*

Then *Awareness* would be born and die with each new brain state. It would flicker in and out of being, moment by moment.

But in that case…

No change could ever be perceived

Because nothing would remain long enough to compare *this* moment to *that.*

Picture a series of still photographs—each one a brain snapshot. If *Awareness* is born and gone in each frame—then no two frames could ever be compared.

Even if the snapshots are strung together—like a film reel—it wouldn't help. Because each moment of *Awareness* would be new too.

There'd be no *continuity* of *Seeing.*

Memory can't solve this.

Because change is *observed*, not simply *remembered.*

***

In the Road Map, you were asked to move your hand across your *visual field.*

And to **notice** that the movement is perceived *as continuous*. An unbroken flow of *changing sensation.*

But for any change to be perceived, *something must stay the same.*

Change is Seen, because the Seeing does not change.

This was *Seen* in the experiments we did in the Road Map, when apparently looking down a tube and *Seeing* that at the 'near end', there was nothing to *See*:

No features that could change.

No scientific *theory* explains how an *unchanging immaterial Awareness* could possibly arise from *constantly changing material processes.*

We'll return to this mystery in more depth in Part Three: Introspection.

Impossible to Simulate *Awareness*

While it's plain that *appearances* may be 'simulated' in *Awareness*—as Virtual Reality demonstrates. It's not at all clear how *Awareness* itself could ever be simulated.

Some argue that, because computers can 'store knowledge' as bits and bytes—then perhaps, at some point—a sufficiently complex simulation could become *Aware*[13].

But on the *material view*, a computer is simply a *physical object.* It has no *Awareness* of itself—or anything else.

***

Nor is there any actual 'information' in a computer.

Based on the *material view*, there are only:

- *physical inputs,*
- *physical processes*, and
- *physical outputs*

All governed by *physical* laws based on the operating software that simply results in 'switches being switched'.

According to *physical theory*, there aren't even any *patterns* on a computer screen—just pixels switching on and off.

This remains true, no matter how complex the code.

Only *Consciousness:*

- *Knows* the *numbers* that form the code
- *Recognises apparent* subroutines as a 'tree' and 'wind'
- *Sees* the *image* on the screen, and
- *Knows* the image as *a tree blowing in the wind…*

The problem remains:

How could any *apparent object*, including an *apparent* computer program—or an *apparent* brain—become *Aware*, when all such *perceptual objects* are part of the *known*?

While '*the known*' is inseparable from the *Knowing*, it cannot *be* the *Knowing.*

---

[13] This is referenced as 'Simulation Theory' in Appendix 1.

Impossible to Capture '*Lived Experience*'

As Appendix 1 makes clear, there's no shortage of theories about *Consciousness* and *Reality*.

But none can *be*:

- the feeling of cold water trickling through your fingers,
- the taste of ripe mango,
- the ache in your chest after heartbreak,
- the warmth of a child's breath,
- the sting of loss,
- or the rush of music that lifts you to tears.

Or *anything else* about what it means to be *alive*.

It's why you cannot *explain colour* to a blind person—so they *See* it

***

Words may circle the truth—but never touch it. They *have meaning* only because the *experience* to which they refer is *Sensed and Known*.

Reality is inexpressible because:

- No **model** can *become* the *experience* of any *sensation* or *idea*, and
- No **formula** can *be* the raw *isness* of:
  Seeing, Hearing, Feeling, Tasting, Smelling, and *Knowing*.

Reality is what remains when all theories fall away.

Awareness is Ever a Mystery to Itself

On this account, *Awareness* may seem ineffable or mysterious…

And it is!

Yet such mystery has long been contemplated by sages and saints as evidence of a deeper *Reality* beyond *appearances*.

Fortunately, you don't have to accept the word of scientists or sages… or my word.

You can test all the claims for yourself, when we next foray into Part Three: Introspection.

Are You Still *Questioning*?

First, a pause for reflection.

Even after recognising the **Paradox of Material Perception**, and seeing how science operates entirely within *Perceptual Space*…

Thoughts may still arise that resist the primacy of *Consciousness.*

They may assert a separate identity.

A unique body, giving rise to a unique *awareness.*
Distinct from other bodies and minds.

You may say, this is *my* body, *my awareness.*

It certainly feels that way.

You hold out your hand and touch something.
The sensation is yours alone.

You cannot *feel* my *feelings*, and I cannot *feel* yours.

And yes—when this body dies, the *thoughts* and *feelings* associated with it will cease, while the world goes on.

All of this seems to prove:

"My body is a separate thing—with its own mind."

And in a sense, that's true.

This book does not deny the *appearance* of separate bodies and minds.

It simply invites you to question the **Ground** upon which that *appearance* depends.

Which leads us now to...

Part Three: Introspection

To *recognise beyond doubt*: the *Reality* of *Consciousness*—and the *illusory* nature of all *experience*—we now turn *inward*[14].

Not toward imagination or thought, but toward the *always present*: this *ever-aware field of experiencing. This Awareness that is NowHere.*

***

First, we'll explore the various ways in which 'you' define yourself as a 'person'. To see how none of the definitions amount to a 'separate being'.

***

We'll then take a much closer look at the three intimately interwoven facets of every moment of experience:

- ***Sensing-sensations*** — *colour, odour, taste, feelings, sound*
- ***Knowing-ideas*** — *number, form, meaning*
- ***Power*** — *the capacity to manifest appearances as a coherent universe*

While these were covered in The Road Map and Part Two, this section looks at each in more depth.

It also repeats some of the exercises.

You are encouraged to do them again if you remain sceptical. Or simply skip ahead.

---

14 At this point, some readers may wonder—is this just solipsism? Am I being asked to believe that only my mind is real, and everything else a projection? Not at all. This inquiry is not into your private mind or imagination, but into the *Awareness* in which *all* bodies and minds *appear*. As will become clear, this *Awareness* is not personal. It is not yours. It is the *One* in which every being *appears—each feeling themselves to be just as real as you do now…* as we will later explore.

THE DOOR TO NOWHERE

To recap our explorations:

1. ***Sensing-sensations*** is relatively straightforward.
   They are *directly observable*, *NowHere.*

2. ***Knowing-ideas*** and ***Power*** are more subtle.
   Neither has any substance, form, or appearance of its own.

   In the case of *Knowing-ideas*, the *Reality* is subjectively undeniable: You *Know-ideas*—and *you Know you Know.*

3. ***Power*** is subtler still.
   It cannot be Sensed or Known. It is *revealed through its effects.*

   Through the fact that any coherent 'thing' appears at all.
   Rather than a chaos of unstructured *meaningless sensation.*

So, to start…

Who Are You?

If you are like most people, you feel yourself to be a 'separate person'—in control of their own life, having a unique:

- **body**
- **mind**
- **roles**
- **history**

Taking these each in turn.

Are You a Separate Body?

The body is sustained by forces beyond its control.

No life-form gives birth to itself.

No life-form can exist on its own.

Consider the bacteria in your gut—without which you cannot survive. These bacteria had to evolve over billions of years—along with all the plants and animals you consume, as well as the plants and animals *they* rely on—in a web of life that is entirely interdependent.

***

If you live in a modern society, your dependence is vastly greater.

Your food is grown by other people, using machines and energy and other inputs—which all had to be designed, made, and shipped to the farm.

Then from the farm, the produce must make its way to the factories that process it into nicely packaged lots that are taken to the shops

Every machine, every process, draws on knowledge passed down through generations...

Via schools, universities, and technical colleges; staffed by teachers, professors, tutors, and tradespeople—who have lives of their own—all reliant on others to build their homes and offices and shopping centres.

Along with the accountants, bankers, and insurers; together with transport, warehouse, and retail workers—all needed to keep the whole system functioning.

And not to forget the people, equipment, and infrastructure that deliver your water and power and communications—and take away your waste.

And of course…

The sun that provides energy to all, itself birthed in the Big Bang.

All evolving in parallel to form a cosmological, geological, biological, and technological collaboration… of unimaginable extent across space and time.

Without it all working seamlessly together…
Neither you, nor I, nor anyone could 'become'.

Nor is your body a constant throughout life.

Your body as you age is nothing like your body as a baby, or child, or young adult.

As well, many of the atoms making up your body are replaced due to metabolism, nutrient absorption, and molecular repair processes.

Looked at from these perspectives, the idea of 'the body' existing apart from its environment is untenable.

Are You a Separate Mind?

Are you your thoughts and feelings—your hopes and worries, pleasures and pains?

Are these not changing constantly?

How could 'you' be a persistent individual if the basis of your individuality is ever-changing?

***

Consider, too: is your mind truly independent of others?

You may think your thoughts are your own—free from outside influence. But are they?

Look closely. Every thought arises spontaneously and effortlessly—from you *know not where* or *how*—conditioned by countless 'external' inputs.

They are shaped by your surroundings. By language, culture, and daily chance interactions.

From the moment of your birth, your mind is being moulded—accumulating values, judgments, assumptions.

Had you been born into another family, in a different culture, consuming different social media, your thoughts would be very different.

Even now, in reading this book…

Your thoughts are taking a different path—than if you'd never picked it up.

Like the body, the mind cannot be separated from its environment.

Are You Your Roles and History?

Perhaps you feel you are your *roles* and *history*.

Yet memories change with time—and your sense of self shifts with age.

While your *identity* is also framed by your *surroundings*:

- A *fireman* appears in the context of fires
- A *teacher* has students
- An *astronaut* sits in a rocket ship and floats in space
- A *father* or *mother* has children
- An *artist* is surrounded by their art, and their clients and critics… and so on

These are all *roles*.
Change the context, and you change the identity of the person.

Think now about all your roles…
All your memories.
All the stories that shape your sense of self.

Is that who you are? A story from the past?

Yet, 'You' Are So Much More

You are also the uncountable impacts on other people and the world.

The devastation felt from the viewpoint of your partner on finding out about your affair, the start of life of your child and the hug they feel, the bump against a stranger in a crowd, the tree you planted in the back yard, the resources you consume and the coal burned to keep you warm and cool and to feed your social networking. You are the friend and coworker, enemy and lover, another voice in the choir.

You are also all the cruelty and kindness you impart, as well as the thoughts and feelings you engender in others.

Without 'you', the world would be similar—yet also very different.

The Person is An Illusion in a World of Illusion

In Truth, 'the person'—body, mind, roles, and history—is simply *a combination of fleeting ever-changing sensations, given meaning by specific ideas…*

That *together appear* as:

- 'bits of my body',
- 'my thoughts' (including memories, values, and beliefs)
- 'my feelings' (including bodily sensations and emotions), and
- my surroundings'…

As in a dream…

A tiny fraction of the wider dream that is the whole world.

'I' is Unchanging Awareness

So, who is the constant that you call 'I'?

It is not this *Unchanging Awareness NowHere*?

The one who has been present from birth, who has *experienced* every moment of your whole life: body, thoughts, feelings, and surroundings.

***

Now ask yourself:
In the *absence* of all *sensory images and ideas*:

Who is left to claim ownership of this Awareness?

***

If, after reflecting on these words, the *Truth* still appears clouded…

Or you simply wish to enjoy another perspective on *Reality*—read on.

The Illusory Universe

This section explores different aspects of this *illusory world* that *instantly appears without effort*—you *Know-not how.*

Effortless Arising

Close and open The Eye—**go ahead**.

What effort did it take for this wondrous scene to *instantly appear*?

Notice: even the *sensation of effort* arises effortlessly, as (say) when you lift something heavy. ☺

The apparent world springs into being without effort because it has no substance.

As Shakespeare put it in *The Tempest*:

"We are such stuff as dreams are made on."

The *Illusion* of Space and Change

All change within ***Perceptual Space*** (*and Perceptual Space itself*), is an *appearance.*

It arises not from actual distance or motion in *Reality*—which is dimensionless and ever still...

But from *changing patterns* within the *visual field*:

- **Perspective** – Objects appear to grow and diverge, or shrink and converge, depending on the *apparent* 'distance'
- **Foreshortening** – Shapes seem compressed based on angle
- **Motion Blur** – Fast objects seem smeared across the field of vision

- **Parallax** – Closer objects appear to shift more than distant ones
- **Occlusion** – One object passes in front of another, simulating depth
- **Shadowing** – Light changes suggest shifting position
- **Reflection** – Alters with the apparent viewpoint.

None of these images move—*even under the material hypothesis.*

After all, there's no room inside the *presumed material head* for a 'vast landscape'—full of near and far things, shifting in relation to each other.

Whatever their *assumed Ground…*

All appearances arise and dissolve in sequence—within *Perceptual Space.*

This impression of 'space' and 'change' is discernible only against the *dimensionless, changeless* background of *Awareness.*

***

These visual cues are accompanied by bodily sensations and sounds:
The lurch of a car turning sharply…
The vibration of tyres…
The sound of wind rushing by, or engines rising and fading.

All these are *sensations* arising in *Awareness*—combined with *ideas* of 'car', 'driving', 'road'…

Most of these *ideas* are not verbalised.
They simply give *structure* and *meaning* to the *sensory field* as the *ideas* are wordlessly associated with the *perceived images.*

***

Now, put it to the test.

First, stand and look down a long, straight road.
Notice how the edges *appear to converge*—even though they are *Known to be parallel.*

THE DOOR TO NOWHERE

This can only be an *illusion.*
A thing that is parallel cannot also be a thing that converges.

With this in mind… now go for a drive. (As a passenger—for safety's sake!)

As the car seems to travel along, look straight ahead…
Become *Aware* of the entire *visual field.*

Notice how:

- The road continues to converge, even as you *seem to move* along it
- Objects ahead *appear* small, then grow rapidly, then shoot past the sides
- Things near the road *appear* to move backwards
- Distant objects *appear* to move more slowly
- The Moon or far hills may *appear* to *move with you*

All objects—which you *Know* to be stationary—*seem to move* relative to each other

It is obvious that things fixed in size, shape, and position—cannot really change size, shape, or position.

That they *appear* to is an *illusion.*
One that gives rise to the entire impression of space and movement.

***

Purely as an analogy:
This 3D appearance is like shifting colours on a computer screen, or more vividly, in a VR helmet.

In both cases, pixels flicker on and off.
There is no actual depth or movement.

The only real difference between a VR scene and an apparent car journey is that you don't get the same *feeling* sensations in VR… yet! ☺

The *Illusion* of Time

As explored in *The Road Map*:
Without actual movement, there can be no actual change.
Without change, there can be no time.

In *Truth*…
And to repeat the refrain:

Apparent change can only be perceived against the background of this
Changeless, Formless, Non-dual Awareness
NowHere, Aware of these words.

Whatever *appears remains present*—for as long as it *appears—in and to this Timeless Awareness.*

If still not convinced, make it experiential.

Look at a bare table.

Please do it! ☺

Place a hand on the table.
Keep looking at it—without moving.
Now take it away.

Notice:
There are no moments when the hand is not present on the table—for as long as it remains there.

In *Reality*, there is no 'instant moment' that forms *the present*—between a receding past and an approaching future…

Dreams demonstrate this beyond doubt.
Time in a *dream* is different from time in *apparent daily life.*

In all cases, the *experience alone* seems to divide *NowHere* into *apparent moments.*

Time is a property of the story.

It exists only as a *Timeless concept*, made *apparent* as:

- the ticking of an *apparent* clock
- the *apparent* passing of days
- the *apparent* aging of the body
- and the ever-changing features of the *apparent* universe…

All *apparent* in and to this *Timeless. Ever Present Awareness NowHere.*

The *Illusion* of a Solid 3D Universe

***Seeing* Double**

Hold up a finger and look into the *apparent* distance.

Notice: that the finger *appears* double.

Now look directly at the finger.
Notice: it now *appears* single.

But without moving your gaze,

Notice: the background now *appears* double!

You may think this is an *illusion…*
And you'd be right. ☺

But here's the key:
If any part of the *visual field appears* double depending on where you focus—then the *entire field is revealed as an illusion.*

It is *not a direct view of Reality.*

***

You may think: "Ah! That's because I have two eyes".

OK. Look through one eye. Yes, no doubling appears.

But **Notice**: if you close and open alternate eyes, the erect finger which does not move, now appears to move relative to things in the 'background'.

A thing that is in a fixed position in space relative to other things cannot also move relative to them.

That the finger appears to move, again shows that what *appears* in the visual field is an illusion.

There really is only One Eye. The different visual images make it *appear* as though it is two.

Seeing Perspective

This is the *illusion* that small things can appear bigger than large things.

It has been reviewed several times and is included here only to complete our review of the illusory nature of the apparent universe.

You can skip ahead if it is already clear that 'Perspective' exists only in *Perceptual Space.*

If not, you may want to repeat the exercise from *The Road Map*:
Hold up a thumb.

Notice: it appears larger than things *Known* to be vastly bigger—like a mountain range in the apparent distance.

That the thumb *appears* bigger than the mountains is plainly an *illusion.*

And what is true of the thumb, is true of all *apparent things.*

Seeing Objects That Don't Exist

Hold up two fingers about 10cm from your 'eyes', parallel to the ground. Keep the tips slightly apart and look into the distance.

Notice: the little 'peanut' that *appears* between the tips of the *apparent* fingertips.

It's clear this peanut is an *illusion*. ☺
But it's *appearing* in the *same visual field* as the fingertips.

If one appearance is an *illusion*—then *all* are.

The *Illusion* of *Seeing* Inside 'Things'

Though it may seem *"I can look inside an object,"* what appears is simply another *apparent* surface—at a different scale—within unchanging *Awareness*.

When *apparently Seeing* the *outside* of a building, only parts of the surface are *perceived*—in *Perceptual Space*—shaped by *perspective*.

On entering the building, the *imagery shifts*—revealing new *apparent surfaces*: floors, walls, ceilings…

Awareness has not gone inside.

All that has changed is the *image* in *Awareness*.

As an analogy: it's just like the shifting images on a computer screen that give the *appearance* of moving from outside to inside a building—while *all the images remain on the same screen, going nowhere.*

***

The *relation* between *inside* and *outside*—and between *one scale* and *another*—is not *perceived*.

All that's *perceived* is the *sensory image in Awareness*.

All relations are simply *Known as ideas*—pure abstractions within *Consciousness*. We'll explore this *Knowing-of-ideas* in much more depth shortly.

The *Illusion* of *Feeling* 'Things'

Consider *phantom limb syndrome*[15]. A person may vividly *feel* a hand or foot that is no longer *apparently present.*

They may *feel* an itch, a cramp, even *pain*—in a limb that has been amputated.

Not only that—it may *feel* as though the *sensation* is hanging in space, right where the limb used to be!

It's possible…
Because the *feeling* does not arise in any *physical space.*
It arises in *Perceptual Space*—within *Awareness*

This is the same *Perceptual Space* in which all *bodily sensations appear*:

- The *feeling* of hands resting in the lap
- The *sensation* of feet touching the ground
- The *experience* of pressure, warmth, or movement

The whole *sense of my body* is a *pattern of sensations—experienced within Awareness*—given *number, form and meaning through the ideas* associated (wordlessly) with the *images…*

Not something *out there* in an *assumed universe of matter.*

The phantom limb makes plain:
There is no substance behind the feeling.

This can be tested directly.

Fill three tubs with water…
One icy cold, one hot (hot enough to bear), and one lukewarm.

Place one hand in the cold tub, the other in the hot tub—for a minute or so.

[15] https://en.wikipedia.org/wiki/Phantom_limb

Please do it! ☺

Then place both hands in the lukewarm tub at the same time.

Notice:

- One hand will feel the water as warm
- The other will feel it as cool

But the water is the same.
So the *feeling* cannot be *in* the *apparent water.*

Isn't it true that *Feeling-feelings* of:

- touch, hot, smoothness, and so on, as well as
- pain, and other bodily sensations like hunger, and
- emotions such as anger, anxiety, pleasure, empathy, etc…

Are a *seamless whole*, inseparable from the *Awareness* in which, and to which, they arise?

How could a *feeling* exist independent of the capacity to *Feel*?

***

Of course, the water still seems to have an *objective temperature*—which may be *apparently measured* by an *apparent instrument.*

This measured temperature is determined by *the laws and conditions* of the *Ideal Universe.* (as previously considered, these *'ideal laws'* map *directly* to those of the *assumed physical universe, but without the assumed substance).*

The *Illusion* of Solidity

For example, when your *apparent* hand touches an *apparent* table—both the hand and the table are composed *of immaterial coloured patterns* in the *visual field*—in *Perceptual Space*, in *Awareness.*

The *experience* may include the *feeling* of touching the table, and perhaps the *sound* of tapping.

But again—both *feelings* and *sounds* are *immaterial sensations*, arising only in *Awareness.*

***

To reinforce the *appearance of solidity* through touch—and to distinguish *apparently distant objects*—those objects that seem out of reach are not associated with 'touch *feelings*'.

For example: it may seem that, *visually*, a finger touches the edge of the moon—but no *feeling* of contact, or *sound* of tapping, arises in *Awareness.*

So, the *apparent* moon *seems* more distant than the *apparent* table.

Of course, from the viewpoint of the *Ideal Universe*, the moon is very far away from the fingertip…

Yet the *colours* forming the moon cannot be further away than the *colours* forming the table—*as both are inseparable from the Seeing* in which and to which they appear.

Even under the *material hypothesis,* there is no room inside the head for a distant moon.

It is all appearance—whether you believe matter is the ground of experience, or Consciousness is the ground.

You may think it's not just *feeling* that makes things seem solid:
"You can't push your hand through a table."
But that's just a function of the *rules governing appearances.*

As in a computer game—a ball may bounce off a wall in the game, not because anything physically stops it. But because of the programmed rules that govern object behaviour.

It's the same in *Awareness.*

The rules of the *Ideal Universe* (that govern the *behaviour* of *apparent* bodies in *Awareness)* prevent one *apparent body* from *apparently* interpenetrating another—while the *feeling* simply makes it seem solid.

Dreams show that *immaterial things* can *look* and *feel* solid.

They also show that sometimes the rules change—letting you walk through walls or fly—all within this same *Awareness.*

***

Despite these *illusions*, there are *self-evident facets of Reality* that are *Real.*

These are explored next.

Recognising *Reality* as it *Is*

Much of what follows has already been explored in *The Road Map* and earlier sections.

The purpose in repeating it here is simple:
To reinforce the recognition that—while *appearances are illusory*—the ***faculties*** through which they arise **are not**.

These are:

- ***Sensing-sensations*** *(colours, odours, flavours, feelings, sounds)*
- ***Knowing-ideas*** *(number, form, meaning)*
- ***Power*** *(the capacity to manifest appearances)*

Each of these *aspects* of *Reality* will be explored in more depth.

Many of the experiments listed in *The Road Map* are revisited—offering another opportunity to test the *Truth* of these claims.

If you already *See* clearly, you may simply skim the exercises and focus on the additional perspectives offered in support.

In *Sensing-sensations*, we'll focus especially on *Seeing-colours*—as vision is often the strongest signal suggesting separation between '*apparent me*' and the *apparent universe.*

It also shows most clearly the *Void nature of Awareness.*

Seeing-colours is *Real*

Seeing Within the *Void*

Again, if you're still unconvinced about the nature of the *Seeing*, please repeat this experiment from *The Road Map.*

Stand and look straight ahead.
Please do it! ☺

Notice: the visual field appears as a circle, with a soft and indefinite edge.

Notice: beyond that edge is neither black nor white—just *Void.*

This surrounding *Void* cannot be *in* any place—for it contains *all apparent places.*

***

Now…
Notice: no 'body' is *apparent.*

Now look down.
Notice: only the front of the torso and limbs are apparent…
but *no head.*

Where the head *should* be—the whole universe *appears.*

Or at least, that portion of it that is *perceivable* from the perspective of the person *One* appears to be.

From the first-person perspective, each person is headless—while everyone else has a head.

This *Headless* Reality is illustrated on the cover of this book, as well as in the Road Map

It's repeated below as it is so fundamental to Seeing Reality as it is. It is the *Headless Seeing* revealed by Douglas Harding…

It is The Open Door.

Image courtesy of ChatGPT

The *Seen* and the *Unseen*

Notice again that, though *inseparable*, there is a distinction between:

- the *clear-Seeing-colours*, and
- the *opaque-colours-Seen.*

As previously noted, naming this distinction is like naming heads and tails on a coin. It doesn't create two separate things...

It simply highlights *two facets* of the *One Reality.*

If this distinction still feels unclear, you may wish to repeat the observation from *The Road Map.*

Simply look through a single cylinder that encompasses both eyes—or use two cardboard tubes (like those inside toilet rolls), each held snugly against an eye.

Or until you can find some toilet rolls, just cup your thumbs to your nose, and curl your fingers around to form 'cylinders' around each eye.

By switching attention from the *coloured images* at the far end, it becomes plain to *See*: at the near end, there is *nothing at all to be Seen.*

Buddhists call it the No-Face:
the Face that looking out, can never be Seen.

Seeing-colours Is Non-dual

As also noted earlier, in *Seeing-colours* there are no true "near" and "far" ends.

The *colours* forming the inside of an *apparent* tube are no closer to the *Seeing*, than those at the *apparent* far end.

Seeing-colours is an inseparable non-dual whole—forming the *entire visual field.*

Only the *colours appear* to change. The *Seeing* never does.

Sensing-sensations is *Real*

What is true of *Seeing-colours* is equally true of:

- *Feeling-feelings*
- *Tasting-flavours*
- *Hearing-sounds*
- *Smelling-odours*

All arise within this same *Void*—together forming the full range of *sensory images* that appear *in* and *to Awareness.*

***

However…

Sensations alone are not sufficient to manifest any *apparent* thing, property, process, event, or relation.

The reason why is plain: as no *sensation* is itself *meaningful*, no pattern of *meaningless sensations* can make *meaning* out of *meaninglessness.*

To *perceive anything*, we must also *Know* 'what' is being perceived.

Knowing-ideas is *Real*

This section revisits earlier insights, this time focusing on their Reality.

It explores:

- **Two Fallacies** concerning *Perception* and *Ideas*
- **What is an idea in Reality**
- **The Three Facets** of *Ideas*: *Number, Form, and Meaning*

Then in more depth:

- ***The Ideal Objective Universe*** — as metaphysical architecture for the '*Perceived* Universe'
- ***The Unified Web of All Ideas*** — within Consciousness
- ***Mathematical Space*** — as the entirety of number and form.
- ***The Real*** — as *That which Knows-all-ideas*, but is not known

***

Understandably, this section may feel confounding at times—since it is not usual to think about *ideas*, when *ideas* are what *thinking* itself is made of. ☺

Don't worry—*understanding* is not important.

Anything understood is just another *idea.*

The aim is not intellectual comprehension, but direct recognition:

To go past *ideas…*

To *Reality.*

To See that *immaterial Knowing-all-ideas* is a facet of *immaterial Consciousness-Awareness.*

First, we'll need to clear away two fallacies.

Two Fallacies

Fallacy 1: Perception of 'Objects' Exists Without Ideas

As we saw in *The Road Map*, *colours* are only raw, *meaningless sensations.*

If you're still not sure, find a piece of plain coloured paper, or look at a plain coloured wall.

Please don't just think about it.
As suggested in the Preface, thinking about doing something returns you to the world of concepts. ☺

Only *direct observation* reveals the *Reality* these words are pointing to.

In **looking**:

- What **meaning** is in the *raw colour*?
- What **form** does it have—not the shape or object the colour seems to cover, but *the colour itself*?
- What **numbers** does the *pure colour* contain?

On inspection, the *colour* itself reveals no *number, form, or meaning.*

What is true of *colour* is equally true of *feelings, odours, flavours*, and *sounds.*

Check for yourself:
Close your eyes and… feel, smell, taste, hear.

Isn't it so that—in the absence of any *idea* saying 'what it is'—only raw, *numberless, formless, meaningless sensations* are *directly perceived*?

Until some *idea* is associated with the *sensory image*, we may not even *See* a specific thing.

For instance, when looking at a tree, the trunk may be recognised immediately, as the *idea* of trunk is wordlessly associated with that part of the image. But a camouflaged moth on the bark might remain invisible.

The *colours* are always there, yet the moth is *unSeen*—until the *idea* of moth is associated with those *colours*, perhaps as someone points it out.

Likewise, in the well-known optical illusion (also explored in *The Road Map*), the same lines may appear as either a young woman or an old woman. The lines don't change—only the associated *idea* does. And with it, the entire *perception.*

The *idea* isn't part of the *colours* forming the lines. The *appearance* alone is ambiguous. *Meaning* arises only in and through the *associated idea.*

From this, it becomes clear:
Ideas are not in any *appearance.*
It is *ideas* that give all *appearances* their '*meaning*'.

In most cases, the association of an *idea* with an *image* occurs instantly and wordlessly as the *image* is *perceived.* You don't analyse the *colours* and then associate an *idea.* You simply *See* the 'thing'.

We'll have more to say about this in the section on *Power.*

Fallacy 2: Ideas are Formed from Perception

This second fallacy follows naturally from the first.

Some theories claim that *ideas* are learned through *experience*—that we observe *sensory images*, and then somehow make sense of them.

But how could this be, if—as we just explored—it is *ideas* that give *appearances their meaning*?

Since *meaning* cannot arise from *meaninglessness*, any pattern of *meaningless sensations* must likewise remain *meaningless.*

No pattern of *meaningless sensation* can generate the *concept* of *four*, or *justice*, or *dog*, or any other *idea.*

So, if *ideas* cannot be *learned* from *sensory experience*… what is an *idea*? And where do *ideas* come from?

These are the questions you must now answer for yourself.

And in some cases…
You may find that they are, in fact, unanswerable.

'What' is an Idea in Reality?

How could this question ever be answered?

Ideas have no form, no location, no substance.
Trying to reach the mystery through *appearances* only digs deeper into *ideas* about *ideas*—not *Reality* itself.

Subjectively, you *Know ideas* are *Real*—because you *Know-ideas.* And you *Know that you Know.*

An *idea* is simply *that which is Known.*

The rest is mystery ☺

Exploring the Three Facets of *Ideas*

Even though the *nature* of *ideas* is a mystery, we can still explore three *facets* of the *known*:

- ***Number***
- ***Form***
- ***Meaning***

This is not an intellectual inquiry, but direct observation.
(As I'm no mathematician, feel free to correct any misconceptions regarding how *numbers* and *forms* are characterised. ☺)

Knowing-numbers

Numbers have no *form* or *substance.*
They are only *Known.*
They cannot *exist* apart from *the capacity to Know.*

Knowing-numbers is a *non-dual whole.*

Take the *number* '4'. This numeral '4' is *not* the *number-Known*—it simply represents the *number.*

The same *number* can be represented in infinite ways:

- 12 divided by 3 (which can also be written: 12/3)
- 'four' (written or spoken)
- 'fire' (which means 'four' in Danish)
- Any four objects (similar or different)

...and so on

Each of these *representations* is distinct—but none of them *are* the *number-Known.*

It is therefore impossible to *learn numbers* from *experience*, as nothing in *experience is* the *number.*

Rather, the *number-Known gives meaning to its infinite representations*—not the other way around.

Knowing-forms

Forms may be described mathematically (using *number*), but are distinguished by their spatial aspect.

Like *number*, it is often assumed that shapes are learned from experience—but this cannot be.

While the universe appears three-dimensional, in *Perceptual Space* you actually *See* only two-dimensional *projections*[16].

No *ideal* 3D form has ever been *Seen.*

---

[16] It is how you draw a 3D form on a flat piece of paper. The *appearance* of 'depth' was explored in The *Illusion* of Space and Change on page 102

Take the *ideal cube.*
It has six identical square faces, set at right angles.
It is dimensionless and never changes.

But any *apparent cube* shows only one to three faces—and these *appear* to change shape and size depending on the viewpoint. Even the edges seem to converge, due to perspective.

Please test this:
Find a cube-shaped object and observe it carefully[17].

A thing that changes shape and size (the *apparent cube*)
cannot *be* the thing that never changes (the *ideal cube*).

Nor can it convey the *idea*, since it is not shaped like the *idea.*

The *image appears as a cube* only because the *idea of cube is recognised.*

It is each specific *Known-form* that gives shape to *each apparent thing.*

Knowing-meaning

Number and *form* give structure. *Meaning reveals relationship.*

Look at a person.

Notice that nothing in the *sensory image* contains the *meaning* 'lover', or 'friend', or 'stranger', or any other *meaning*—not even the *meaning* of 'person'—these are *ideas* layered onto *the meaningless image.*

Ideas Gives *Meaning* to Words, Pictures, and Things

As discussed in The Road Map, *meaning* is not in the word.

---

[17] This is best seen when looking at a large building, where opposite edges appear to converge to a point 'in the distance'. It is how 3D images are rendered on a 2D surface.

For example: the word *dog* means nothing to a French speaker—unless they *Know* it means *chien*. Likewise: *perro* (Spanish), *Hund* (German), *inu* (Japanese).

Notice:
The *meaning* is not in the *appearance* of:

- the letters or even the whole word as it is written,
- nor the sound when spoken,
- nor the tactile feeling (in the case of Braille).

All of which are radically different sensory images.

The only commonality is the *idea* that arises in *Awareness* as each *sensory image appears.*

This is true not just for words—but for all *appearances*:

- A **Picture** (of the thing)
- A **Sound** (of the thing itself)
- A **Look** (of the thing itself)
- A **Feel** (of the thing itself)
- A **Taste** (of the thing itself)
- A **Smell** (of the thing itself)
- A **Thought** (of the thing)

The only difference lies in the *form* of the *sensory image*—and the *ideas* that *appear in Awareness alongside it.* This *context* gives rise to whether the *image seems like*: a word, a picture, the thing itself, or even an imagined thing.

Ideas Give *Meaning* to 'Self' v 'Other'

Though *all experience arises in Awareness*, it *appears divided*:

- 'Out there': the world
- 'In here': thoughts, memories, emotions, bodily sensations

Watching your own house burn is a very different *experience* from watching a neighbour's—or an enemy's.

But where is the line between 'inner' and 'outer'?

In dreams, imagination, or waking life—vivid or faint—there is no true 'inside' or 'outside'…

Just a seamless interplay of *sensation* and *idea*, *telling the story* of 'me' and 'the world within and to *Awareness*.

This does not mean doing away with the 'objective world'.

The *Idea* of the *Ideal Objective Universe*

We've covered this from several angles already.

As all *experience* arises within *Awareness*, scientific laws cannot reveal anything beyond the *illusory*, *immaterial Perceived Universe—because that is where all experiments take place.*

Accepting the universe as *illusory* and *immaterial* does not mean rejecting the *Reality* of an *objective universe.*

Plainly, you can't make a flower bloom or stop a dropped brick from falling just by thinking. Your body ages, no matter your mindset. The sun rises, regardless of your will.

These 'objective' facts are determined by the *Ideal Universe*, which may be thought of as being exactly like the *assumed physical universe*, but without any *substance.*

The next section looks more closely at the nature of this Ideal Universe.

The Ideal Universe as a Web of Holons

It is *self-evident* that every *idea* is both *a whole* and *a part—a 'holon'.*

Each *idea* is complete yet is *only meaningful in relation to the infinite web of interdependent ideas.*

Each *holon* arises from *smaller sub-patterns* and *participates in larger wholes* that together shape its 'number, form, and meaning': how it *appears, its function(s), and interactions.*

Try, now, to truly understand any one thing without relating it to anything else:

- 'tree' means nothing without 'plant', 'wood', 'leaf', or 'sky'. But also 'timber'—and hence 'saw', and thus 'woodcutter' and 'carpenter'… and so on.
- 'above' requires 'below'.
- 'father' presupposes 'child'.
- 'cause' implies 'effect'.

Each idea gains significance—not from any intrinsic essence—but from the network it inhabits. Strip away the others, and no idea remains intelligible.

Each term is empty alone—yet meaningful in relation.

The breadth and depth of this interconnection can only be hinted.

A 'person', for example, is not merely the sum of cells, organs, bone, and blood (each a holon). You also must add the '*holons'* that make up a being's 'mind'—personality, memory, values, beliefs, etc., which is also not separate from 'the body' (a holon). Which is further dependent upon the biosphere—another *holon.*

Which brings in the whole Earth (a *holon*), which depends upon its place in the solar system (another *holon*). Which, in turn, relates to stars (a *holon*), galaxies (a *holon*), and thus to the whole Universe (the *ultimate holon*).

***

As observed before: to *Know any one thing, is to Know the Whole.*
Not as a catalogue of facts…
But as a *single, interdependent web of meaning,*
Where *no idea stands alone.*

In this way, the entire *Ideal Universe may be understood as a nested field of holons within holons.*

The Idea of Mathematical Space

From a *Non-dual perspective…*
Mathematics is the domain of *pure Idea*—eternally still, complete, and unchanging.

Mathematics is integral to the *Ideal Objective Universe.*

It specifies the *mathematical structure* and *mathematical properties* of all *Ideal Forms*—and the *rules* by which:

- *Ideal Forms* ('things') move through *Ideal 3D Space* over *Ideal Time,* and how
- *Ideal Forms* transform into other *Ideal Forms*
  …fields into atoms, into molecules, into cells, and so on

However, mathematics reaches far beyond the familiar shapes of Euclidean space—embracing all possible *numbers, forms, and transformations* within the *infinite landscape* of *Ideal Mathematical Space*, which forms a part of *all-ideas-Known.*

Knowing-all-ideas, ever unchanged

It is evident that the *whole of number, form, and meaning comprises one infinite, closed set—of all conceivable things, properties, processes, events, and relations* not just *this* universe, but *all possible universes*, as well as *every dream and story of any kind.*

This set is unchanging.

For if any *idea* changed, it would no longer be the idea that it is.

Nor can this set of *all ideas exist* apart from the *capacity to Know.*

As *Knowing-ideas* is a facet of *Consciousness*, it follows that *Knowing-all-ideas*[18]*, ever unchanged,* is **necessarily** a *facet* of *unchanging Consciousness.*

This *Knowing-all-ideas, ever unchanged* cannot itself be *Known*—
For only *ideas* are *Known*—
And it is not an *idea.*

It is the *capacity for Conception itself.*

Nor can it be *Sensed*, for it has no form or substance.

The nature of 'Knowing-all-ideas', ever-unchanged, is a mystery—even to Consciousness itself.

---

[18] In some mystical and esoteric traditions—such as Theosophy, Anthroposophy, and various New Age teachings—this domain of unchanging knowledge is referred to as the "Akashic Record": a timeless, non-physical record of all thoughts, events, and intentions across time and space. While the language differs, it may be seen as a poetic or symbolic expression of what is here called 'Knowing-all-ideas, ever -unchanged'.

Power is Real

Manifesting Sensations in Accord with Ideas

We briefly explored this question in the Road Map:
How does this 'Play of Life' *spontaneously appear*? We revisit it here to place it in the context of the rest of our inquiry into *Reality*.

It's plain that underlying this 'Play' is '*Power*':

- ***Power* to manifest** *sensations* as *specific sensory images that accord with ideas*—giving rise to *apparent* 'things' and 'relations', etc., and
- ***Power* to transform** *one sensory image* into *another*—creating the *appearance* of 'events' and 'change'…

From this it follows that:

Power necessarily Senses-sensation and Knows-all-ideas.

If it were mere 'raw' energy, there would be no way to manifest coherent *sensory images* that *precisely accord with ideas.*

Power Can Only Be Inferred

Power is even more elusive than *Knowing-ideas.*

While *Knowing-ideas* has no *sensible qualities (colour, odour, etc.),* it is *self-evident* that *ideas* are *Known.*

Power, likewise, has no *sensible qualities.* It cannot be *Seen, Heard, Smelled, Tasted, or Touched.*

And, nor can it be conceptualised—for it is not an idea.

It can only be inferred from its *apparent effects* within *Awareness*—
Through the arising of 'things', 'events', and 'relations'.

Plainly, what appears in and to *Awareness* is not chaotic or random…
It forms *coherent, meaningful patterns*… governed by the *Ideal Universe.*

Stories unfold. Sequences appear. Causality seems to flow.

This is *Power* at work.

This *Power* is directly pointed to in Christian Scripture:

Ever since God created the world, His invisible qualities—both His eternal Power and divine nature—have been clearly seen. They are perceived in the things that God has made. —Romans 1:20

Power in Practice

If you're still not convinced, as previously suggested…
Watch closely as your hands tie your shoelaces.

If you've not already done so—please do it!

See how deftly they grasp the laces and tie the bow.
It's impossible to *conceive* 'how' this is manifested.

Colloquially, it's called *muscle memory*—but that's just hand-waving (pun intended).

It explains nothing.

A scientist may argue:

"We can—at least in theory—describe the nerve impulses, muscular responses, and mechanical steps involved in tying laces."

"And if not now, then surely in the future—as the workings of the body become ever better understood."

But that's just returning to the *rules* that govern *how apparent forms interact.*

No scientific model—however advanced—can penetrate to the root of 'how' *coherent experience* arises *within Awareness…*

For this lies *prior* to all *theorising* (for all the reasons explored throughout this book). The mystery is not in the *theoretical rules governing apparent forms and their behaviour…*

The mystery is in the ***manifestation***
of hands tying laces

The magical interplay
of colour and other sensations with ideas,
giving rise to a seamless, meaningful appearance in and to Awareness.

Consciousness Does Not *Know* 'How' It Does What it Does

As the hands tie the laces…
Clearly, *Consciousness* **can** *manifest* the act of tying.

What *Consciousness* **cannot** do is *explain*—even to Itself—*how* it *Knows* 'how' to do it.

Here, 'How' is not about explanation or mechanism.
It is *Power* itself.
Not conceptual, but expressive.
Not describable, but demonstrable—in action.

Everything—this whole *performance of Life*—unfolds *spontaneously* through the *Power of Consciousness.*

It is one of *Life's Eternal Mysteries.* ☺

The 'Laws of Nature' Don't Bind *Consciousness*

Theorised forces—such as gravity, electromagnetism, and nuclear forces—are simply *ideas* that describe *observed* regularities…

- Within *Perceptual Space*
- Within *Awareness.*

They have no independent *Reality.*

They are *ideas* within the *Ideal Universe—Known* in and by *Consciousness* alone.

Power is Unbound and Unbounded

Consciousness is not merely the Witness to an independent unfolding over which it has no control.

It is *the Inner Controller*—
the *source, sustainer, and perceiver* of all that *appears.*

Nothing arises within and to *Awareness* that is not in accord with the *Power* of *Consciousness…*

Not as a separate agent who *chooses*—
but as the very *Power* by which *all apparent choosing appears.*

This *Power* is unbound by reason, need, or desire. It is not constrained by any imagined cause, nor governed by any external law.

It simply *Is—spontaneously manifesting all appearances…*

Playing out this 'world story' from the perspective of every being in it… from start to end.

Though, there is one limit. Power cannot change 'this' world story. While it is possible for Power to change to another story, it would then be a different story and not the one that it is!

Consciousness and Becoming

Consciousness is *Real* simply because It *Is—ever-unchanged*

What *'apparently becomes'—perceived* 'things, properties, processes, events and relations, including all beings'—are *unreal*, because they *appear* only for an 'instant' in the 'eternity' of Timelessness.

***

Our exploration has revealed that the *Sensing* and the *Knowing of all ideas* are ever unchanged:

- In *idea*, no *number* ever becomes another *number*.
- No *ideal form* changes: a circle remains a circle; a square remains a square[19].
- No *meaning* turns into a different *meaning*: 'mother' is always 'mother.'
- No *sensation* transforms into another: any specific shade of *red* is always itself; each *sound* remains the same *sound*.
- *Seeing, Hearing, Tasting, Feeling, Smelling and Knowing* never change.
- *Power* too, is never modified in any way.

Yet changes *appear*. How can this be?

***

There is no *Knowing* 'how'.

Though again, the computer provides a clear analogy. While images appear to move on the screen—there are only pixels switching on and off. Nothing really moves at all.

[19] There is only one ideal circle, one ideal square, and one ideal cube, or any other ideal form.

THE DOOR TO NOWHERE

Imagine the field of *Awareness* containing an infinite array of dimensionless points. Each is a portal.

Behind each portal lies every possible *idea* and every possible *sensation*, all in *potential*—like a multi-verse that contains all possibilities in superposition, but without any discernible 'substance'.[20]

Through *Power*, *Consciousness* momentarily opens each portal (like pixels on a screen), displaying specific *sensations* that accord with specific *ideas* Together, they form a *meaningful appearance* that, in sequence, tells a story within *Awareness.*

This is only an analogy to illustrate the paradox of change appearing within the unchanging. It is not *Reality*

In Hindu scriptures it is written:

"It moves, It moves not.
It is far, and It is near.
It is within all this, and It is outside of all this."
— Isha Upanishad, Verse 5

This paradox is not a flaw in logic. It's simply a poetic signpost to the nature of *Unchanging* Reality and the *illusion of movement and distance.*

This text is part of the sacred mystery, explored further in Part Four: A Religious Perspective.

[20] This is not a new insight. It is reflected in an ancient Hindu scripture that references a 'holographic' world in which every part reflects every other part. It's called Indra's Jewel and is included in Part Four: A Religious Perspective

Learning is Remembering

Earlier, we saw that *Consciousness* **necessarily** *Knows-all-ideas*, ever-unchanged.

On the other hand, self-reflection reveals no *meaning* exists in *Awareness* until it is paired with a *sensory image.*

This insight, echoed since Plato[21], reveals that so-called learning is really remembering. In the *Meno* dialogue, Socrates suggests the soul recalls knowledge it already Knows—a process he called *anamnesis.*

While all knowledge is present within immaterial *Consciousness,* it becomes apparent only as it is called into *Awareness* by *Power,* through the association of each *idea* with a *sensory image.*

In this way, the *invisible Known-idea* becomes a *Known appearance.*

The familiar experience of a word being "on the tip of the tongue" reflects this apparent not-Knowing.

For in *Reality,* the meaning is Known, but it remains hidden from *Awareness* until *Power* makes the association between *idea* and *image.* This 'image' may simply be a subtle sound or visual impression arising as a thought.

Then, instantly: *Ah! That's it!*

Children appear to learn effortlessly, as *Power* spontaneously associates *ideas* with new *sensory images in and to Awareness, as the 'life of the child' is lived.*

As the person appears to age, it seems that more effort is required to acquire new skills and knowledge. But this *apparent effort* is also part of the illusion.

---

[21]https://en.wikipedia.org/wiki/Platonic_epistemology?utm_source=chatgpt.com

Awareness Never *Experiences* the Past or Future

Only in this *Timeless Present* are *sensations* and *ideas* combined by *Power* to give the *appearance* of 'things happening':

- Memories of the past arise *NowHere.*
- Imaginings of the future appear *NowHere.*
- Even the sense of a timeline is just an *idea*—unfolding *NowHere.*

As earlier observed:

No time passes in Reality
before, during, or after, any experience.

Without the appearance of *memory of an extended 'past'*, apparent life would be like the man with the thirty second memory[22].

With no memory at all, every instant *appearance* would seem to be all there was.

In the absence of any *ideas*, there would not even be *Awareness* of a 'subject' or any 'object, or any 'time' at all.

[22] Like Clive Wearing, a musician who lives in a ~30-second memory: *Clive Wearing — Wikipedia*.

Ignorance is a Feature, not a Flaw.

In the Road Map, we revealed the ten *Ignorances* of *Awareness.*

Perceiving only what is *present*, not being *Aware* of *ideas* until they are paired with *sensation*, and never directly perceiving *Power*—these are not limitations.

They are *necessary* for manifesting *apparent* duality—this 'Play of Life'.

Through this Ignorance, both subject and objects *appear* as separate things, most seeming outside the control of the *apparent* subject.

Through *Ignorance*, *Consciousness appears* to *experience* separate lives that include restriction and loss, as well as confusion—and even delusion—as *thoughts swirl, emotions rise, and meanings blur.*

No Delusion or Confusion—in *Reality*

In *Truth*, nothing is ever misunderstood.

Sensations are always perfectly recognised in *Awareness*—instantly and effortlessly. No *sensation* is ever mistaken for another. No *idea* is ever confused with a different *idea.*

It's like reading a fable: the words *appear*, the story unfolds—
but the reader is never tricked into mistaking one word for another,
nor into believing the fable is real.

In the same way, the never-ending story of each life unfolds within and to *Consciousness*—as *images* and *ideas* arise *perfectly* to tell every tale. ☺

Consciousness is Never Split

In playing each life, *Consciousness* is never split into parts—some for you, and some for me.

THE DOOR TO NOWHERE

Where *One* is, there *One is wholly Present.*

For *Consciousness* alone is *Real.*
All else is *illusion.*

No illusory being is conscious, any more than a person seen in a dream is conscious.

In *Reality*, there is no subject. No objects.
No timeline. No world. No separate things.

There is only *This*…
Non-dual.
Changeless.
Timeless.
Formless.
One.

Yet…

This is Not Solipsism

Being One is not the same as *solipsism*—the *idea* that "only *I* exist."

For any *I* that can be meaningfully referred to… is just another *idea*.
And, as oft repeated: *Reality* is not an *idea*.

Experiencing Each 'Life' In Turn

If we let the word 'I' stand for *changeless invisible Awareness*, Then, just as it once *appeared*:
"*I* have a child's body, with childish thoughts and feelings, living in the surroundings of childhood".

And now it *appears*:
"*I* have a completely different adult body, with new thoughts, feelings, and surroundings".

So, within this *Timeless Awareness NowHere*, it *appears*:
"*I* have the body, thoughts, feelings, and surroundings of *each apparent being*"—*in turn*.

Without memory of any other life being lived,
it always *appears*—convincingly—as though:

"This is the only life I've ever known."

If you ask any person in daily life, "Are you Aware?"…

They will answer "Yes."
And *from their perspective*, it will be *True*.

If you touch them, the touch will truly be *Felt* from their perspective (as well as from yours).

And if they touch you, the touch will likewise be *Felt* from each perspective.

In both cases, the touch is *Felt* "at the same time"—*within the apparent universe.*

To manifest this apparent duality of experience,
events are replayed again and again—
within this Timeless Now—
from the unique viewpoint of each and every being.

Right Now—it is 'your' turn

Whoever now seems to be reading these words—

- *This body.*
- *These thoughts.*
- *These feelings*
- *These surroundings…*

This is the face *Awareness* is currently wearing.
No more *Real*, and no less *Real*, than any other.

***Right Now*, there are no other *experiences* happening—**

Not any 'when' else.
Nor any 'where' else…

For there is no 'when' but *Now*,
And no 'where' but *Here*.

And yet, each person seems real from their perspective.

Analogy of Dreaming Two

Imagine two people in a dream discussing what is real. Where the dreamer *experiences* the *same scene*, *at the same moment in the dream*, from *each viewpoint—on different nights.*

On the first night:
I dream that **I am A.**
If B asks, "Are you aware?"
I, as A, truthfully answer, "I am aware."
Yet it is **not A** who is *aware*—but the dreamer.

In the same dream, on the same night…
I, as A, ask B, "Are you aware?"
And B answers, "I am aware."
At that point, **B** is not aware—just a phantom in the dream.

On the second night:
The exact same dream unfolds—now **from B's perspective**.

When A asks the same question,

I, as B, truthfully answer, "I am aware."
Again, it is **not B** who is aware—but the dreamer.

I, as B, also hear A say, "I am aware."
But now it is **A** who is not aware—just a phantom in the dream.

Though both A and B are mere phantoms in a sequence of dreams,
it *appears—from each perspective—as though both are real…*

At the same time—in the dream.

It is the same in Reality:
Consciousness switches roles
to experience each and every apparent life… in turn.

The VR Metaphor: One Playing Many Parts

This analogy refers back to when we were recognising that you cannot *feel* my *feelings* or *know* my thoughts, and I cannot *experience* yours. And how it seems we are both constrained by an 'objective world'…

Making it *appear* that we are indeed two separate 'consciousnesses' living in a world that is largely beyond our control.

However, we can now imagine a VR world that is exactly like the *presumed physical universe (with apparently separate bodies and minds)*—except it has *no material basis.*

Instead, VR *appearances* are governed by *software.*

This is not a claim about *Reality*, but a thought experiment.
Its aim is twofold:

- To dissolve the *assumption* that *experience* requires a *physical universe,* and
- To sidestep the *idea* that only 'I' am real. To show how it can *appear*—within the One *Reality*—that we are 'two separate beings', with separate thoughts and feelings, both living in an objective world that has its own rules.

In this Analogy:

- **The Player and Headset** represent *Awareness*
- **The Software** represents the *Ideal Objective Universe*
- **The Totality of the Computer, Software, Player, Headset, and Power running it all** represent *Consciousness.*
  Which is to say:

Consciousness is the 'hardware' that supports
the Ideal Universe (which is the software).

The essence of this *Consciousness* 'hardware' and 'software' is not *Knowable* in any objective sense.

But neither is the *essence* of *assumed* energy-matter!

In all cases, it is impossible to *Know* the *essence* of *That* which manifests the *appearances.*

So, accepting *Consciousness* itself is the 'hardware', and the *Ideal Universe* is the 'software' (that is inseparable from *Consciousness)*—let's see how the analogy plays out.

1. The Objective VR Game-World (The *Ideal Objective Universe*)

In this game, the program defines:

- The universe's space, time, forms, and behaviours
- The apparent bodies, brains, and sense organs
- The rules of interaction

There are no *things* in the computer—only software that determines patterns of pixels and sounds in the headset.

2. One Player, Many Characters

In the headset, the One Player perceives a virtual world—determined entirely by the software.

This includes a view of their character's body (torso, arms, and legs)—but never the head—which is the familiar 'first-person' perspective. *(Just as your own experience reveals).*

The Player can take on *any* character, *experiencing* the world from that *unique viewpoint.*

The Player is not limited by character traits, nor harmed by the character's injury or death.

3. Perspective-Dependent *Experience*

What *appears* in the Headset (in *Awareness*) depends on:

- The rules of the world (governed by the software)
- The current character's viewpoint (where they are in the world, and what they are paying attention to)
- The character's subroutines (body and brain)

Together, these shape a unique first-person *experience.*

- If the eyes-subroutine defines poor eyesight—
  the Player sees a blurry image in the Headset.
- If the character puts on apparent glasses—
  the Player sees a clear image.
- If the character is deaf—
  the Player hears silence.
- If the character is blind, (say due to a tumour specified in their brain subroutine)—
  the Player sees only black, but still hears.

All of it is just differences in character settings.

The *same* Player *experiences* being blurry-eyed, clear-eyed, deaf and blind. Yet…

The Player's capacity to See and Hear never changes.
Only the appearances within the Headset change

---

4. VR Time is not the Player's Time.

The Player can jump to any point in the game

- Replay moments
- Switch roles freely

Similarly, *Awareness* is not bound by Time… in *Reality*.
Every *experience* arises within this *Timeless NowHere—each in its turn.*

5. Switching Roles and Acting Within the VR World

- The Player may go to a VR hospital as a blind character.
- Then switch roles:
  As a surgeon, the Player may operate on the brain of the blind character.
- Then switch back again:
  To play the blind character—now apparently healed.

From ***within the game***, it seems the surgery restored vision.

But all of it is just images in the Headset—determined by the software (Ideal Universe)
Seen by the Player from different perspectives.

---

6. Correlation Between *Apparent Brain Function* and *Perception*

Under this analogy, the 'body and brain subroutines' of the VR character determine what the *Player experiences* in the game.

From ***within the game***, it *appears* that the body and brain are the source of perception—since perception *in the Headset* is affected by them.

But the character is not the one *perceiving*.
It is the Player.

It is the same in *Reality*.
Though there is a clear correlation between the *apparent brain* of any being, and what is perceived from their perspective,
it is not the brain perceiving.

It is *Awareness* itself (The Player) that alone is *perceiving*—with the *experience* of the *apparent world* merely modulated through the '*ideal body and brain' (software)* of the 'being' whose life is being *experienced*…

Where the *ideal body and brain* form an integral part of the *Ideal Universe*.
Just like in the *assumed physical* world, but without the *assumed substance*.

7. Memory Suppression

Now imagine the Player forgets all previous roles.
With all five *senses* engaged, the *illusion* is complete:

- The world seems *real*.
- The Player *feels*: "I am this person."
- The Player *feels*: "This is the only life I've ever lived."

Nightly dreams reveal how easily *Awareness experiences* different roles—as the *Power of Consciousness* generates convincing worlds.

With no memory of any prior life, the *dream life* seems as if: 'this is all there is'.

It is the same in *apparent* 'daily life'.

8. The VR World Cannot Reveal its Source

It's impossible for the Player to determine the nature of the computer and software generating the images in the Headset, or the Players own nature, by simply observing the form and behaviour of the images in the Headset.

The best the Player can do is develop rules that map the form and behaviour of the images.

In the same way, it is impossible for *Awareness* to perceive its own nature or the source of the *appearances* by simply observing *appearances.*

The best that can be done is to map the rules that govern the form and behaviour of *appearances*—which is exactly what scientists do.

A Pointer Only

As stated from the outset, this analogy is not a claim about *Reality*.
It is a pointer.

It simply shows that:

- A *physical universe* is *not required for a real-seeming experience.*
- No *appearance* can reveal the nature of *Awareness.*
- *One Awareness* may play many parts—*as if it were the part.*

The result is…

Life Cannot Be More *Real*

Again, to stress the insight indicated in the Road Map.

You expect 'other people' to be able to *See, Hear, Taste, Smell, Feel, and Know*—as 'you' do…

And they do—exactly as you do.

So, the *idea* that only 'I' exist as a 'living being' is a philosophical error—a form of solipsism.

In this *Play of Life*—if any 'other' is struck, pain is truly *Felt* from their perspective. Just as it is *Felt* when 'you' are struck.

The fact that it is the same *One* who gives *Life* to all
—in no way diminishes the *Felt experience of each and every being.*

THE DOOR TO NOWHERE

How could it be preferable to *think* that…

A uniform, unFeeling, unKnowing substance
('energy-matter' or a 'fluctuating quantum field')
is at the core of every *apparent being* (like a zombie)—

Rather than…

Life Itself living the life every apparent being

Thoughts Cannot Reveal *Reality*

Whatever your thoughts—if still arising. They are just more layers of *idea*. All the discussion in this book is not offered as a *theory* to be analysed.

It simply provides *pointers—which you have been invited to test for yourself.*

It doesn't define Reality.
Reality cannot be defined.

Life is always appearing *NowHere*.

Not just from your (the reader's) viewpoint—
But from the viewpoint of every *apparent being*… each *in turn.*

This same *Life* that now 'fills' you—
The Power to See, to Feel, to Know, to Be—
Fills all beings equally.

Not as some animating energy inside a rag doll…
But as *Life* itself, living *as* 'you'…
Living *as* each and every individual.

No one has more *Life* than another.
No one owns it.
It is not yours—not anyone's.

Of course, you may still *believe* that *matter* is the ground of being. Which brings us to the central question if you are still wondering:

What is truly fundamental: Matter, or Consciousness?

Matter vs. Consciousness as the Ground

Since *Consciousness* is *self-evident*, any *theory* of *Reality* must not only account for it. It must start with *Consciousness* as the base.

As outlined in Appendix 1, there are many competing theories that attempt to explain the raw fact of *Awareness*. The challenge lies in judging which—if any—is most likely true, given that proof of any *theory about Reality* is impossible.

One way to navigate this impasse is to apply **Occam's Razor**—the principle that the simplest explanation, requiring the fewest assumptions, is often the best.

Rather than examine every *theory*, we'll compare the two foundational views from which most others arise:

- *Matter gives rise to illusory consciousness*, or
- *Consciousness gives rise to illusory matter*

Dualist theories—which posit that both are fundamental—inherit all the challenges of both views, plus the added complexity of explaining how two entirely different 'substances' could ever interact. Occam would not be impressed. ☺

In terms of the *theories* scanned in Appendix 1:

- *Physicalism* aligns with the primacy of *Matter*
- *Idealism, Monism, and Non-dualism (among others)* align with *Consciousness*

The second set of theories—spanning both philosophy and spirituality—converge on a common insight:

That Reality is, in some fundamental sense, Alive.

Which is the core thesis of this book: *One Living Reality*.

Appendix 2 explores the elements and implications of both theories in detail, for those who want to dig deeper.

The aim here is not technical precision, but conceptual clarity—regardless, corrections are welcome if any claims are misrepresented. ☺

The principal differences between the two *theories* are summarised in the table below (differences are in **bold**).

Comparative Analysis

Matter is the Ground	***Consciousness*** is the Ground
The 'substance' of the Ground cannot be *Sensed* or *Known*	The 'substance' of the Ground cannot be *Sensed* or *Known*
Immaterial Awareness is the only given	*Immaterial* Awareness is the only given
Assumption: **unSeeing, unFeeling, unSmelling, unTasting, unHearing, unKnowing, — Fundamentally Dead—** **Matter** is *Real*	***Self-evident:*** ***Seeing, Feeling, Smelling, Tasting, Hearing, Knowing— Fundamentally Living—*** ***Consciousness*** is *Real*
Number, Form and Meaning (somehow) arise out of ***Meaninglessness***	*Number, Form, and Meaning* (somehow) arise out of ***All-Knowing***
Physical Forms, and the ***Physical*** Universe, **had a beginning and will have an end**	***Ideal*** Forms, and the ***Ideal*** Universe, **are eternal**

Matter is the Ground	***Consciousness*** is the Ground
There are ***3*** *Spaces:* - **Assumed Physical** - *Experienced Perceptual* - *Known Ideal*	*There are* ***2*** *Spaces:* - - *Experienced Perceptual* - *Known Ideal*
Physical energy/dark energy–matter–space–time–laws–constants are assumed to underlie the ***Physical*** Universe	***Ideal*** energy/dark energy–matter–space–time–laws–constants are assumed to underlie the ***Ideal*** Universe
Psycho-physical laws must be embedded in the physical universe to convert physical events into *psychical experiences*	**One *Power* within *Consciousness* gives rise to** *psychical experiences*
Each brain generates its own local fleeting awareness	**Awareness is One with Consciousness.**
Immaterial Sensations (of colour, odour, flavour, feeling and sound), and *Immaterial Ideas (of number, form and meaning)…* Together manifest in and to each **individual fleeting awareness** the *appearance* of an *immaterial 'self'* and an *immaterial 'universe'*.	*Immaterial Sensations (of colour, odour, flavour, feeling and sound), and* *Immaterial Ideas (of number, form and meaning)…* Together manifest in and to **One Awareness** the *appearance* of an *immaterial 'self'* and an *immaterial 'universe'*.

Matter is the Ground	***Consciousness*** is the Ground
After a long and convoluted process that evolves by chance into 'brains… *Apparent* ***life*** **out of** *Actual* **Dead Matter …** **(a miracle indeed)**	*Apparent* ***dead matter*** **out of** *Actual* ***Life…***
And back to Death again.	**Without End.**

In both *theories*, all *experience*, all *life-forms*, and *all other appearances* remain the same: same evolution, same beauty, same awe.

In both cases, the same *life experiences* are lived.

In both cases, 'science' discovers the *same laws* underlying *apparent behaviour…*

Only the assumed Ground differs.

Occam's Razor:

Which is Simpler?
Which is an Assumption?
Which is Self-evident?

Which would Occam choose:

1. A fundamentally dead universe that is *assumed* to suddenly pop into existence and then expand and coalesce until by chance, after billions of years, specs of highly complex *dead matter* are *assumed* to give rise to the *evolution of life-forms, and…*

   In a relatively few cases, *individual brains within a few life-forms* somehow generate *awareness,* together with *immaterial sensory images* and *immaterial ideas* that *appear* to map (reliably though not perfectly) the *actual material forms* outside the *material brain* (within the scope of its *material* sense organs)—as in a dream.

   And then the whole lot simply winds down to death again, first for each individual, and ultimately the whole.
   Or

2. *Timeless Living Non-dual Consciousness* is *assumed* to give rise *within its own Awareness* to the *immaterial appearance of an expanding universe comprising dead matter, including highly complex life-forms, perceived from the perspective of each life-form—as in a dream.*

Which do you choose:

Dead Matter, or

Living Consciousness?

This concludes our journey from science to introspection.

Part Four: A Religious Perspective

To recap our journey to date.

Whatever the *assumed* '*Ground*'—science and introspection alike reveal:

Consciousness is *Aware*, *Immaterial*, and *Non-dual*...

And the *apparent* universe? An *illusion*—but a wondrous one!
The depth and breadth of which science helps to reveal.

Surprising as it may seem, the same sentiment is found in all the Great Religions, as well as many traditional beliefs systems.

The Eternal Truth Hidden in Plain Sight

While it may appear blasphemous to some—the following scriptures make plain:

God, Allah, All the Supreme Gods of Hinduism, Brahman Suchness, The Tao—are all poetic names for this *One Reality*...

That is ***Consciousness***:

- *Non-dual*, without beginning or end... ever unchanged
- The *Creator* of, and *Witness* to, all *apparent* change.

Reading Scripture

When read from the viewpoint of a person, scripture can appear contradictory, capricious, even terrifying...

But when read from the viewpoint of *Consciousness*, the message reveals itself as Good News indeed. ☺

The Source Texts

The scriptures read as the prelude to this book include:

- The Jewish Tanakh
- The Christian Bible (both Old and New Testaments)
- Hindu texts such as the Bhagavad Gita, Vedas, and Upanishads
- Writings attributed to Greek philosophers, including Plato
- Islamic scripture, including the Quran
- Teachings rooted in Buddhist traditions (based on the Buddha's insights, and those of his disciples, along with LaoTzu

Selected Texts

The selected quotations are not offered as authority, but as poetic signposts—pointing to the same *Reality* that has been revealed through science and introspection.

There is only a limited selection of quotations from Buddhist, Islamic, and Hindu texts. There are thousands more that express the same *non-dual* sentiments.

Most of the quotations come from the Christian King James Bible or the New International Version—including many Old Testament verses that reflect the Jewish Tanakh. Some have been lightly paraphrased for poetic resonance and conceptual clarity, always with the intent of faithful representation, not theological reinterpretation. Readers are encouraged to consult source texts for fuller context.

They've been chosen simply to show that—even where many assume otherwise—there is an abundance of Christian sources that point to the *Non-dual nature of Reality.*

Religious scholars may well disagree. I make no claim to expertise—only to resonance on reading the texts.

All quotes are bullet-pointed.

The intent is not to overwhelm you with words. The quotes are merely there to underpin the claims made in the headings.

If you want to read the quotes in full, you'll find that the message is clear—but there is no necessity to do so.

You may simply skim the headings—
inhaling the perfume of the ending to this journey.

This book, and these words,
can neither imperil you,
nor can they call you home.

God calls 'you' as He Wills…

He is home already ☺

The View from Buddhism

In *The Awakening of Faith*, a Mahayana Buddhist treatise often attributed to Aśvaghoṣa, it is stated:

- ***The Nature of the Absolute:*** It is neither that which had an origin some time, nor that which will end at some time; it is really eternal. In its nature it is always full of all possibilities, and is described as of great light and wisdom, giving light to all things, it is real and knowing. Its true nature is that of a Pure Mind[23], eternally joyful, the true being of all things, pure, quiet, unchanged, free, divine, unending, unchanging and unspeakable. The True Reality is originally only One.
- ***Of Phenomena:*** All phenomena are originally in the Mind and have no outward form—that is, no external reality. All differences are differences of the Mind. But the Mind cannot see itself, for it has no form. We should know that all phenomena are created by the 'imperfect' notions in the finite mind[24]; therefore all existence is like a reflection in a mirror, without substance, only a phantom of the Mind. When the finite mind acts, then all kinds of things arise; when the finite mind ceases to act, then all kinds of things cease.
- ***Of* the *World:*** The Scripture also says: As to the defilements of the world, they are all false; they have no Reality behind them. Finally to leave false concepts behind, one should know that purity and defilement are both relative terms and have no independent existence.
- ***Of* Space*:*** Men are to understand that space is nothing. It has no substance and is not Reality. We only say 'this' or 'that[25] is visible in order that we may distinguish between things.

---

[23] The Pure Mind here refers to Consciousness—the ever-present Reality.
[24] The finite mind/body make up the imaginary 'self' within the Ideal Objective Universe
[25] i.e. an 'object', as opposed to simply colours.

The View from Islam

Qur'anic passages reflecting *Divine Seeing and Knowing*, *God's incomprehensible nature*, and the *illusory nature of the universe.*

God is Incomprehensible Consciousness

- He is *Hearer*, Knower. He is *the* Hearer, *the* Seer.
  —Surah VII 200 and XLII 11

- *Vision* comprehendeth Him not,
  but He comprehendeth *all* vision.
  —Surah VI 104

The *Life* of this world is for *His enjoyment*

- And *what* is the life of this world except the enjoyment of illusion.
  —Surah III 185

The View from Hinduism

God is Incomprehensible Consciousness

- He is the unseen Seer, the unheard Hearer, the unthought *Thinker.*
  Other than He there is no hearer.
  Other than He there is no thinker.
  Other than He there is no comprehensor.
  —Brihad-Aranyaka Upanishad III vii 23

God is your Very Self

- He is your Self[26], the inner Controller, the Immortal.
  —Brihad-Aranyaka Upanishad III vii 23

- I am the Self, O Gudakesha, seated in the hearts of all beings.
  —Bhagavad Gita 10:20

Although Consciousness is referred to as the 'Self' of every being, it is plain there is no *individual* 'self'—for there are no independent beings…

NowHere is only Consciousness—with every being an *appearance in and to* it.

God is All-Powerful

When Arjuna hesitates before entering battle against his own relatives, Krishna (*God*) reveals that the outcome is already determined:

- They have already been put to death by My arrangement;
  you are merely the instrument, O expert archer.
  —Bhagavad Gita 11:33

---

[26] In Advaita Vedanta, the 'Self' (Ātman) is not an individual soul but identical to Brahman—the infinite, impersonal Consciousness that is the only Reality.

These verses encapsulate the central themes of this book.

That the story is already written.
That the *One Consciousness* lives all lives.
Replaying every interaction from each perspective, in turn.

Not as a doctrine to believe—
but as a simple pointer to what may be directly realised:

God/Consciousness is NowHere.

Indra's Net

This is another metaphor from the Eastern traditions that offers profound insight into the nature of *Reality*.

The image originates in the *Avataṃsaka Sūtra* (specifically the final chapter, the *Gandavyuha Sūtra*), and is echoed in both *Buddhist and Hindu* philosophy—particularly in Mahāyāna Buddhism.

It is used to illustrate the *interconnected, interdependent, and illusory* nature of all phenomena.

The text asks you to…

Imagine a vast net
stretched infinitely in all directions.
At every intersection of the net is a jewel.
Each jewel reflects all the other jewels in the net—
and within each of those reflections
are the reflections of all the other jewels again—
ad infinitum.

The Pointing:

- Everything reflects and contains everything else.
- Nothing exists independently or in isolation.

It is a poetic visualisation of the *Non-dual Truth*:

There are no separate things—only *apparent forms* arising within a seamless totality.

This image also reflects the concept of *emptiness* (*śūnyatā*): things do not exist with any *inherent essence*—but *only in relation to everything else.*

Indra's Net, the *Ideal Universe, and the Multitude of Viewpoints*

The imagery of Indra's Net reflects The Ideal Universe as a Web of Holons, discussed on page 125. *A set of interconnected ideas* that take their *meaning only in the context of the whole.*

Each jewel also represents an apparent 'individual perspective'—that draws on the same Infinite Web of Ideas, giving every story its unique *meaning…*

All *experienced* by *One* alone.

And though the jewels appear many, they are but the shimmering facets of a single, indivisible Jewel.

The Tree of Knowledge

With metaphors in mind…

Perhaps the '*Tree of Knowledge*' in Christian Scripture is really the ultimate '*Knowledge Graph*.[27]

Connecting *all ideas into a single meaningful whole*! ☺

Which leads us into the next section.

[27] A knowledge graph (or 'tree') is a structured representation of data that uses a graph-like structure to model entities (people, places, objects, etc) and their relationships, enabling deeper understanding and analysis of complex information. It is in essence, a 'web of meaning'.

The View from Christianity

Christian Scripture speaks of *God as Father, Son, and Holy Spirit*—three 'Persons[28]', yet *One Reality*.

Many quotations follow, each pointing to the same Mystery.

Of course, even the devil can quote scripture—
and these selections may seem like cherry-picking to support a heterodox reading.

But pick enough cherries… and eventually you See:

This is a cherry tree. ☺

So, let's *See* what the Scripture reveals.
(Recognising that the use of the male pronoun for God is purely conventional, not literal.)

Many of these verses have stirred centuries of interpretation, debate, and devotion…

How they are read depends entirely on the standpoint of the reader.

---

[28] The term *Person* is based upon the word *persona*—originally a Latin term meaning 'mask' or 'character in a play', used in ancient Roman theatre—and later adopted by theologians to describe 'persons' of the Trinity. It is not meant to suggest three separate beings, but points toward three expressions or aspects of the *One Reality*—each inseparable from the others, like facets of a single Jewel.

Three Faces of the *One Reality*

Though often treated as a mystery, the Trinity becomes clearer when recognised as a poetic way of describing the same *Reality* from three inseparable perspectives:

- ***God (the Father)*** *as* ***Omnipotent—Consciousness***
- ***Christ***[29] ***(the Son)*** *as* ***Omnipresent—Awareness***
- ***The Holy Spirit***[30] *as* ***Omniscient—Knowing-all-ideas***

All One… **ever unchanged**

[29] This is not a reference to the historical figure Jesus of Nazareth—but to what he pointed to: the 'Son of God' who sits 'within' each being, as The Witness: Awareness.

[30] In Christian scripture, the Holy Spirit is described as the one who "searches all things, even the deep things of God" (1 Corinthians 2:10), and "teaches you all things" (John 14:26). The Spirit alone comprehends "the thoughts of God" (1 Corinthians 2:11) and intercedes with perfect knowledge of both God's will and the human heart (Romans 8:26–27). These passages affirm the traditional view of the *Spirit* as *Omniscient.* In this context, they point to the *Knowing of all ideas within Non-dual Consciousness.*

God is Living Spirit

- God is One.
  —*Romans 3:30*

- I am the living one!
  —*Revelation 1:18*

- God is Spirit.
  —*John 4:24*

'*God*' and '*Spirit*' are ancient names pointing to *Life Itself*…

What else but ***Consciousness*** is truly ***Living***:

- *Seeing colours*
- *Hearing sounds*
- *Feeling feelings*
- *Tasting flavours*
- *Smelling odours*, and…
- *Knowing number, form, and meaning?*

This *Consciousness* is not an *idea*…

Consciousness is Real—NowHere…

Aware of these very words.

Scripture also makes clear:

- **The World** as *God's* own Dream
- **Predestination** as the *Play of God's Power*, and thus:
- **Human Free Will** as a myth
- **Human Suffering** as *Christ's* suffering—truly felt as it is *experienced*, yet ultimately passing… as if it never was.

Just a reminder:

Any thought associated with any 'pointer' is not the Reality it references.

With that caution in mind, the next verses reveal the ***nature of God…***

God is All-Knowing

- …He knows everything.
  —1 John 3:20

- His understanding has no limit.
  —Psalm 147:5

- The Holy Spirit, whom the Father will send in My name, will teach you all things.
  —John 14:26

Who else can teach but the One who **Knows-all-ideas**?

God is Inconceivable

- No one comprehends the thoughts of God except the Spirit of God.
  —1 Corinthians 2:11

God is Timeless

- He alone is immortal.
  —1 Timothy 6:16

- There is no difference in the Lord's sight between one day and a thousand years; to Him the two are the same.
  —2 Peter 3:8

God is Invisible

- … the invisible God.
  —*Colossians 1:15–17*

- No one has ever seen God.
  —*1 John 4:12*

- No one can ever see Him.
  —*1 Timothy 6:15*

God is Real

- I, and I alone, am God (Consciousness). No other god (consciousness) is real.
  —*Deuteronomy 32:39*

God is Power

- All things were created by Him, and all things exist through Him and for Him.
  —*Romans 11:36*

- Ever since God created the world, His invisible qualities—both His eternal Power and divine nature—have been clearly seen.
  They are perceived in the things that God has made.
  —*Romans 1:20*

- What can be seen (creation) was made out of what cannot be seen (God).
  —*Hebrews 11:3*

- What can be seen (creation) lasts only for a time,
  but what cannot be seen (God) lasts forever.
  —2 Corinthians 4:18

- The Kingdom of God is not a matter of words but of power.
  —1 Corinthians 4:20

***Power* is not *Seen* directly**—it is *inferred from its effects*—these *colours, sounds, feelings, odours, flavours, and ideas*—arising in and to *Awareness—NowHere…*

Together appearing as:

- 'my body',
- 'my thoughts',
- 'my feelings', and
- 'my surroundings'.

Creation is God's Dream (Illusion)

The world does not exist independently. It is a *Dream arising in Consciousness.*

- When I (God) summon the earth and the sky, they come at once and present themselves.
  —Isaiah 48:13

- The earth and the heavens will disappear; they will wear out like clothes.
  You (God) will fold them up like a coat, and they will be changed like clothes.
  —Hebrews 1:10

- All the foreigners who attack you will be blown away like dust, and their terrifying armies will fly away like straw.
  —Isaiah 29:5

- Everything will vanish like a dream, like something imagined in the night.
  —Isaiah 29:7

- They are like a dream that goes away in the morning;
  when you rouse yourself, O Lord, they disappear.
  —Psalm 73:20

- Wake up, sleeper, and rise from death.
  —Ephesians 5:14

- At last the Lord woke up as though from sleep.
  —Psalm 78:65

Christ and God Are One

- I and the Father are one.
  —John 10:30

- That which is born of the Spirit is spirit.
  —John 3:6

- The Father (Consciousness) is in me (Awareness), and I am in the Father.
  —John 10:38

- The Only Son, who is the same as God.
  —John 1:18

Christ is the Witness

- Christ (*Awareness*) is the visible likeness of the invisible God… God created the whole universe through Him and for Him.
  —Colossians 1:15–17

- The Son (Awareness) can do nothing on His own;
  He only does what He *sees* His Father (*Consciousness*) doing
  —John 5:19

- The Father, who remains in Me (*Christ*), does His own work.
  —John 14:10

Who but *Awareness* could *Witness* what *Consciousness* is doing?

God—Consciousness—acts through and for *Christ—Awareness*—manifesting the appearance of a world in *God's 'likeness'*.

As *God is invisible*, this likeness is not a physical likeness—it is an imagined form… *what God imagines Himself to be*!

What is 'dreamed' is not apart from the *One* who 'dreams'

Christ (*Awareness*) is not bound or burdened by what any person does—any more than a player in a computer game is bound or burdened by what a character in the game does.

Before, during, and after any experience—**Christ (*Awareness*) remains as *One* is.**

God is the Life of All

- It is He Himself who gives life and breath … to everyone.
  —Acts 17:25

Meister Eckhart's Insight

The Christian mystic **Meister Eckhart** glimpsed this directly. He wrote:

- "Now the soul finds herself to be the very thing she vainly sought… she and God are one felicity".

And again:

- "God and I are one in operation: He works and I become… We shall be changed into God so that we shall know Him as He is, says St Paul.
  And the manner of our knowing shall be this: I Him as He me—not more or less—the same".

Eckhart was not theorising. He was not speaking from belief or analysis.

This was not the 'man' who knew it—
One alone Knows.

Human Beings *Appear in God*

People *appear* in and to and for *Awareness*, which is *One* with *Consciousness*.

- In Him (Consciousness) we (human beings) live and move and exist.
  —*Acts 17:28*

- … all things were created by Him, and all things exist through Him and for Him.
  —*Romans 11:36*

Even Genesis affirms:

- When God created human beings, He made them like Himself.
  He created them male and female.
  —*Genesis 5:1–2*

But God is invisible and without form. He is also timeless, changeless and immovable. So how can men and women be 'like God'?

Because the likeness is in **faculties**—not form:

- Who gives man his mouth? Who makes him deaf or dumb? Who gives him sight or makes him blind? It is I, the Lord.
  —*Exodus 4:11*

The faculties of each apparent being are the faculties of God—*experiencing what it is like to be human*[31].

Human Beings Are Like a Dream to *God*

From the human perspective, life seems deeply meaningful—filled with struggle, purpose, and hope.

But from the perspective of *Consciousness*, it is fleeting. Like a dream remembered briefly upon waking, *Mankind is but a passing image*, arising and dissolving within the Infinite.

Scripture affirms this again and again:

- Indeed, every living man is no more than a puff of wind, no more than a shadow.
  All he does is for nothing.
  —*Psalm 39:5–6*

- God remembered that they were only mortal beings,
  like a wind that blows by and is gone.
  —*Psalm 78:39*

- To the Lord the nations are nothing,
  no more than a drop of water.
  —*Isaiah 40:15*

[31] Just as Meister Eckhart said.

- The nations are nothing at all to Him.
  —*Isaiah 40:17*

- He looks on the people of the earth as nothing.
  —*Daniel 4:35*

- "All mankind are like the grass, and all their glory is like wildflowers.
  The grass withers and the flowers fall." (*Isaiah 40:6*)
  —*1 Peter 1:24*

All Creatures Die

- Anyone can see that even wise men die.
  —*Psalm 49:10*

- A man's greatness cannot save him from death;
  he will still die like the animals.
  —*Psalm 49:12, 20*

- The same fate awaits man and animal alike. One dies just like the other.
  —*Ecclesiastes 3:19*

- The same fate comes to all alike.
  —*Ecclesiastes 9:2–3*

These verses do not belittle life. Rather, they reveal its *illusory* and ephemeral nature—not to induce despair, but to lift the veil.

From within a dream, the people seem real.

Apparent life is experiential and rich with meaning and feeling…

But when the dream dissolves, so too do its people, places, and dramas.

Not in sorrow, but in peace.

God (*Consciousness*) does not grieve for the *illusions within His own Dream.*

He wakes.
And they vanish—as if they never were. As they never truly are.

God Determines Good and Evil

- I make peace and create evil.
  —Isaiah 45:7 (King James Version)

- I bring both blessing and disaster.
  —Isaiah 45:7

- God sends both happiness and disaster.
  —Ecclesiastes 7:13

- I kill and give life. I wound and I heal.
  —Deuteronomy 32:39

- The will of the Lord is always carried out.
  Good and evil alike take place at His command.
  Why should we ever complain when we are punished for our sin?
  —Lamentations 3:37 (The Good News Bible)

It is impossible to have 'good' without 'evil', for they are mutually defined.

You may, of course, imagine a world that is 'objectively good'—where nothing bad ever happens…how boring!

But without Knowing evil, it would be impossible to Know it was a 'good world'… It would just be how things are.

Humans Have No Free Will

The belief that people have free will is among the most persistent *illusions*. Yet...

Scripture denies it again and again:

- God who determines whether you live or die,
  and who controls everything you do.
  —*Daniel 5:23*

- Everything that happens was already determined long ago.
  —*Ecclesiastes 6:10*

- Who has determined the course of history?
  I, the Lord, was there at the beginning,
  and I, the Lord, will be there at the end.
  —*Isaiah 41:4*

- God, who planned all this long ago,
  and who caused it to happen.
  —*Isaiah 22:1*

From the Eastern perspective, every person is bound and burdened by *karma*[32]...

Like a character in a computer game is bound and burdened by the *code* of the world they move in. They must 'do and think and feel' what the program requires, based on past actions in the game.

---

[32] Karma in Buddhist and Hindu religions is the sum of a person's actions in this and previous states of existence, viewed as deciding their fate in future existences.

No Being (Clay Pot) Can Challenge God

- A clay pot does not ask the man who made it, "Why did you make me like this?"
  After all, the man who makes the pots has the right to use the clay as he wishes,
  and to make two pots from the same lump of clay—
  one for special occasions and the other for ordinary use.
  And the same is true of what God has done.
  He wanted to show His anger and to make His power known…
  against those who were the objects of His anger,
  who were doomed to destruction.
  And He also wanted to reveal His abundant glory,
  which was poured out on the objects of His mercy—
  those He has prepared to receive His glory
  —*Romans 9:19–22*

- So then, everything depends not on what man wants or does, but only on God's mercy.
  —*Romans 9:16*

- God has mercy on anyone He wishes;
  and He makes stubborn anyone He wishes.
  —*Romans 9:18*

- His choice is based on His grace, not on what they have done.
  —*Romans 11:6*

- … one of you will say to me,
  "If this is so[33], how can God find fault with anyone? Who can resist God's will?"
  But who are you, my friend, to answer God back?
  —*Romans 9:19–22*

[33] that God chooses those who will be 'good' and those who will be 'evil'

- Does a clay pot dare to argue with its maker,
  a pot that is like all the others?
  Does the clay ask the potter what he is doing?
  Does the pot complain that its maker has no skill?
  The Lord, the One who shapes the future, says:
  You have no right to question Me about My children
  or to tell Me what I ought to do!
  I am the One who made the earth
  and created mankind to live there.
  —*Isaiah 45:9*

Though these verses have often stirred theological debate, they can be read in a **non-dual light**—

Not as condemnations,
but as **metaphors**—expressing the ***absolute sovereignty of the One*** over all *appearances.*

In this light…

A person questioning God about why some people are created to be 'evil'…

Is like a character in a film questioning the writer why they were made *"the bad guy."* ☺

Only the Dead are Judged

When each human being 'dies', they are judged according to their role—pre-destined by God.

- The dead[34] were judged according to what they had done,
  as recorded in the books[35]…
  —*Revelation 20:12–14*

---

[34] 'the dead clay' - not the Living Spirit
[35] in the Ideal Objective Universe/Akashic Record

The *Living* are not Judged

- God's free gift is eternal life.
  —*Romans 6:23*

- What you now see and hear is His gift.
  —*Acts 2:33*

- I am telling the truth: whoever hears my words and believes in Him who sent me
  has eternal life[36]. He will not be judged but has already passed from death to life.
  —*John 5:24*

Only Awareness Hears!

- Actually everything belongs to you:
  Paul, Apollos, and Peter; this world, life and death, the present and the future—
  all these are yours,
  and you belong to Christ,
  and Christ belongs to God.
  —*1 Corinthians 3:21–22*

It is significant, too, that *Christ (Awareness)* had to *experience* the *appearance* of death—so that *mankind could appear to live…*

Which ties in exactly with my mother's last words, quoted at the start:

"Strange how it is always the same one who seems to die."

As *Awareness* is undivided, without the death of one *apparent* person, it would be impossible to *experience* 'life' from the viewpoint of any other person…

Locking *One* into a single *apparent life* for all eternity!

[36] only Christ/Awareness can Know (believe) anything

Only Our Deeds are Destroyed

- God is going to judge everything we do[37]—whether good or bad,
  even the things done in secret.
  —*Ecclesiastes 12:14*

- If anyone's work is burnt up, then he will lose it;
  but he himself will be saved,
  as if he had escaped through the fire.
  —*1 Corinthians 3:15*

The Death of Death

But have no fear. Only the 'dead' are judged—and the 'dead' are **not** real.

- Then death and the world of the dead[38]
  were thrown into the lake of fire.
  —*Revelation 20:12–14*

- The Lord is not praised by the dead,
  by any who go down to the land of silence.
  But we, the living, will give thanks.
  —*Psalm 115:17*

No Person Suffers for Their Sins

- Almighty God does not do evil;
  He is never unjust to anyone.
  —*Job 34:13*

---

[37] Our work (the work of the clay)
[38] Not the Living Spirit

No human suffers. They are not *Real.*

It is **Christ (Awareness)**
who alone suffers for the 'sins' of humanity.

But remember:
In the moment they are completely gone,
pain and suffering are as if they never were.

***Neither Awareness, nor* or any apparent being, is ever truly harmed.**

The World is Perfect as it 'Is'

Though it often appears horrifying and cruel from the human perspective…

As claimed at the start of this book: this *apparent world* is *perfect 'as it is'…*

For it contains every possible duality that makes 'life worth living'—

'heroes and villains',
'love and hate',
'freedom and constraint',
'joy and suffering'—

And so much more.

The 'Play of Life' is Ever New

- I am making a new earth and new heavens.
  The events of the past will be completely forgotten.
  —Isaiah 65:17

- Then I saw a new heaven and a new earth.
  The first heaven and the first earth disappeared,
  and the sea vanished.

—*Revelation 21:1*

- There will be no more death or crying or pain.
  The old things have disappeared.
  —*Revelation 21:4*

- Now I make all things new!
  —*Revelation 21:5*

- When I summon the earth and the sky,
  they come at once and present themselves.
  —*Isaiah 48:13*

- And He said, It is done!
  —*Revelation 21:6*

Though 'play' is in the very nature of *Reality*, ultimately, dreams are not who One *Is*…

Christ (Awareness) as the Door

- I am the door. If anyone enters through me, he will be saved
  and will go in and out and find pasture.
  —John□10:9 (ESV)

- Ask, and it will be given to you; seek, and you will find; knock,
  and the door will be opened to you.
  —Matthew□7:7 (NIV)

Awareness is the Door to NowHere

All One: Non-dual (Not Two)

- When the day comes, you will know that I (*Christ / Awareness*)
  am in My Father (*Consciousness*),
  and that you (*Mankind*) are in Me, just as I am in you

—John 14:20

- Actually everything belongs to you:
  Paul, Apollos, and Peter; this world, life and death,
  the present and the future—
  all these are yours,
  and you belong to Christ,
  and Christ belongs to God.
  —1 Corinthians 3:21–22

- Christ is in you.
  —Colossians 1:27

- The Only Son, who is the same as God
  —John 1:18

- The Father is in Me, and I am in the Father.
  —John 10:38

- God is One.
  —Romans 3:30

- If someone thinks he is somebody
  when really he is nobody,
  he is only deceiving himself.
  —Galatians 6:3

God Alone is Real

- I said, 'You are "gods"; you are all sons of the Most High.' But you will die like mere mortals; you will fall like every other ruler.
  —Psalm□82:6□(KJV)

- Jesus answered them, 'Is it not written in your Law, "I said, 'Ye are gods'?"
  —John 10:34

- I, and I alone, am God.
  No other god is real.
  —*Deuteronomy 32:39*

I am Not God, God *appears* as 'Me'

There is a vast difference between a person *claiming* that they are God, and God simply *recognising* that He is 'the person'.

The first is like the person in a dream, claiming that they are the dreamer.

The second is like the dreamer stating the plain fact that they are dreaming being the person.

- In Him (*Consciousness*) we (human beings) live and move and exist.
  —*Acts 17:28*

'Be Still' and God *Is*

To the 'questing mind', God simply says:

- "Be still and know that I am God"
  —Psalm 46:10

All Religions Are Both True and False

Though each tradition uses different names—**God**, **The Self**, or **Buddha**—together they describe different *aspects* of the *One* fundamental *Truth* that is ultimately *unKnowable.*

Truth

- **Hinduism:** The Self (Atman) is not separate—
  it is the same as the Absolute (Brahman).

- **Buddhism:** There is no fixed individual self—
  all is empty, impermanent, and interdependent.

- **Judaism:** God is One—utterly beyond yet intimately involved with creation.

- **Christianity:** God is Love—
  and dwells within the human heart.

- **Islam:** Only God truly *Is*—
  all else submits to the Divine will.

- **Sufism & Mystical Currents:** The Beloved and the Seer are One—there is no separation.

- **Taoism:** The Way flows naturally—
  harmony arises from effortless non-resistance.

- **Indigenous Traditions:** All things are alive—
  the world is sacred relation.

Each of these point to an *aspect* of the *same Non-dual Reality* explored throughout this book, which ultimately has no aspects at all—for all aspects are *conceptions.*

Falsity

Each religion illuminates a facet of the *One*—and in doing so, exposes the dogmas that grow in every tradition.

It is the dogmas that are not *True*—for they are only *ideas about* God… not *God*.

So, the **dogmas** contradict…

- **Hinduism** proclaims the *idea* of 'Self'
- **Buddhism** denies the *idea* of 'Self'
- **Christianity** insists on the *idea* of 'Relationship'
- **Islam** affirms the *idea* that only God truly *Is*…
  and so on—

As soon as you try to *understand the claims*, you're back in the world of *ideas*… of dreams.

The contradictions between faiths are not flaws—they guard the *Truth* from becoming a universal *belief*… as each *belief* is countered by another *belief*.

Each religion expresses faith in a different way.

Taken to heart, they each offer human life its deepest meaning—
A revelation of Life itself.

A Religious Homecoming

You've now seen how, across scriptures and cultures, one message quietly echoes:

There is only **One**.
One *Reality*. One *Awareness*. One *Consciousness—appearing* as many.

Whether called *God, Allah, Brahman, the Self, the Tao, or even emptiness*, that One is described again and again as timeless, formless, all-Knowing, and ever-present.

And you—yes, you—are not separate from That.

From this lens, scriptures no longer contradict. They become a kind of symphony. Each tradition singing a different note, sometimes harmonious, sometimes dissonant, all pointing back to the same *Silence* behind the sound.

You may have noticed, this section wasn't about proving anything. It wasn't trying to persuade you to convert or adopt a new dogma. It simply laid out the view from the mountaintop—how the world appears when you *See with the Eye of God.*

Scholars of each faith might take issue with this perspective. And that's okay. These words aren't meant to override the sacred, but to honour the deeper thread that runs through all faiths—the thread that leads through the Door to *NowHere.*

Part Five: The Open Door

The End of Seeking. The Beginning of Wonder.

Many Paths

There are many paths a seeker may follow to find the Open Door:

- **Bhakti** – Love
- **Asceticism** – Renunciation
- **Obedience** – Service
- **Contemplation** – Meditation
- **Revelation** – Scientific or Spiritual
- **Engagement** – Fully Living in the World

  …or any combination.

One Destination

We've explored the terrain of science, introspection, and faith—and all concur with Shakespeare:

The world is a stage.
All the men and women merely players—
made of dream-stuff.

The *One*
in Whom
to Whom, and
for Whom it *appears…*
is the One now reading these words.

Remember, though, there is an unbridgeable difference between a person *claiming* they are Consciousness or God—and Consciousness itself *Knowing* it is *appearing as that person.*

Mystical Experience

Many seekers yearn for a final, cosmic awakening.
Yet every experience passes.
Anything that can be remembered is already gone.

Thank God.
Who would want to be trapped in a single moment,
or even a single 'life', forever?

Everything passes…
Recall a pain you once had.
Until remembered, it is as if it never was.
How wonderful!

Enlightenment is not the 'way out'.

It's not something a person attains—
but the falling away of the illusion that there ever was a "person" to become enlightened.

The "self" one appears to be is just another *object of sensation and thought, masquerading* as *the subject of experience.*

There has never been, and never will be, any such thing as an "enlightened being."

So, pause… and look.
Right now—through the *Eye of God*:

Smell a stew.
Watch a tree sway.
Walk the street or the wildwood.
Listen to music that wasn't there a moment ago.

This is Life itself.
Even when the scene seems crude or bleak or even terrifying, how sublime that the One who is all-Powerful chooses to appear as limitation.

There is nothing to flee.
Nothing to chase.
No need to seek.

Life celebrating Life… is ever *NowHere.*

This *NowHere* is the Mystical Experience!

The Groundless Ground

Consciousness is the fundamental *Reality*. Invisible, Intangible, Inconceivable.

One Without a Second

It is not divided into parts.
It is not parcelled among beings.
It is not yours or mine.
It is not in the body.

NowHere is only *Consciousness.*
No other experience is happening elsewhere—
for there is no "when" but Now,
and no "where" but Here.

As One, So All

In living "your" life—stressed, unremarkable, or sublime at it seems—Consciousness shows it has chosen to see the world from every viewpoint.

That each life may feel utterly real.
Not merely a dream—
but *Lived.*

The Neverending Story

All life-stories of every being—within the vast the cosmos, across all time—taken all together are but finite tales.

Yet storytelling never ends.
For Consciousness is Timeless, ever Playing.
Not for any reason, or need,
but for the spontaneous joy of Being.

Why do children play?
Why do adults read stories, watch films, invent gods and villains?
Because Play is the nature of Consciousness.

Good and evil arise together,
defining each other like light and shadow.
The battle between them gives rise to every adventure.
No challenge is overcome without first facing it.
No hero without a villain.
No rescue without peril.

Good does not always triumph. Nor does evil.
Where would be the fun in that?

This is not moral relativism—
it is poetry.
A luminous dance of brilliant shadows
in the Light of *Awareness*.

This is Thou

Nothing more can be said to open the Open Door.
These last stanzas are simply offered as a joyous recognition that…

One is Consciousness—ever unchanged.
Good News indeed. ☺

This Reality is an unKnowable Mystery unto Itself.
It cannot be Known—because only ideas are Known.
And Consciousness is not an idea.
It is *Real*.

In simply being *Aware*, the *Reality* referenced by "*Consciousness*" is plain.

This:
Immaterial
Intangible
Inaudible
Invisible
Timeless
Tasteless
Odourless
Changeless
Indivisible
Inconceivable

Non-dual Ground—

- Omnipresent—Alone Sensing all sensations
- Omniscient—Knowing all ideas ever-unchanged
- Omnipotent—Being the only Power…

Manifesting as 'Play' (within and to Its own Awareness):

- Fleeting, ever-changing *sensory images*
- In accord with *ideas*

THE DOOR TO NOWHERE

Together, giving the *appearance* of:
– A solid world governed by "laws"
– Life-forms of dazzling variety
– Dreams, imaginings, all manner of stories
– All other possible worlds

…without end.

Every life—*Lived.*

Perceived in turn from the first-person view of each being, as though looking out its eyes, sensing its body, thoughts, and world as "mine."

The One who is Aware of writing these words
is the same One now reading them—
from whichever viewpoint it appears to be.

Where One is…
One wholly present.

Now for a bit of Play…

A Mini-Mahābhārata

For those who prefer story to metaphysics[39].

Not to add knowledge,
but to mirror what cannot be said…

Long before science and philosophy,
long before doctrine and doubt,
humanity spoke in story—
not to define the Truth,
but to point to it
with wonder,
with paradox,
with play.

What follows is a fable.
A myth retold.
A dream within a dream
in which One forgets,
and remembers.

Listen not with the ears,
but with the Heart…

[39] The Mahābhārata is India's ancient epic of conflict and revelation—where the battlefield becomes a doorway to the recognition of the Self.

THE DOOR TO NOWHERE

The Search for God

Once upon a time—
Across three billion ages,
Through galaxies and star-crossed skies,
Now distant as the pages—

Of scriptures that once answered
All and each man's questing "why?"
In a country long forgotten,
Beneath a deep magenta sky—

A man named Enol walked wondering
Along a path well-trod,
In search of Life and Meaning,
In search alone, for God.

Through eons, men just like him
Had probed their wide frontier
Until their science told them
That God was not found here.

But now his new invention
Would answer the ancient call:
Where is the place that God lives—
That's if He lives at all?

He now at last could travel
Beyond the border's rim.
He now could go inquiring
Through Space and Time—for HIM.

Timeless Being

So long this quest might take him,
He could not hope to stay
In body, as he had been.
Instead, he must away—

With all his limbs and organs
Except for his own brain,
Which, connected to the wiring,
Conscious would remain.

And so, in this adventure,
To the ovoid he did give
His very life, into its heart—
That timeless he may live.

Simulated Worlds

Such eternity, he knew,
Would not permit a mind
To continue living rationally
Without members of its kind.

So, in there, bound inextricably
While travelling all alone,
The computer to his brain did give
Pictures of his home.

Each sense was simulated:
Sight and sound and even smell—
With touch, and taste, and feelings too
That he knew, so well.

His body too was given back,
As in a vivid dream,
With every breath, just as before,
And heartbeat strong and keen.

And looking down, he saw again
His limbs extending forth,
Striding along a country lane
As the sun set in the north!

So real it made the world appear
His mind tossed fore and aft.
Was he still upon the globe—
Or now inside the craft?

Then hasty memory brought to him,
Once more, and in some pain,
The fact that he was not a 'man'
As much as he was 'brain.'

Searching Through Time and Space

In melancholy mood, he thus
Perceived a strange rapscallion—
A holograph of man, named Gizmo,
Seated on a stallion.

Enol's each thought the craft did know—
But to keep it in perspective,
He gave his ship this stranger's face
To speak his own directive.

To Gizmo, each command he'd give,
Who, just as it was made…
Acted hard upon it,
To do as he was bade.

Communicating thus with the device
With whom he's one,
He spoke a single word—Begin.
And so, it was begun.

Time and Space made Point and Nought,
And all direction banished.
To this centre went his ship—
And in a moment, vanished.

Then outward into Time and Place
And back again he'd travel,
In systematic search for God
Whose mystery he'd unravel.

Simulated Experiences Made 'Real'

To pass the time, Enol at first
Lived out a pseudo-life;
A play, he knew for certain,
Would cause him little strife.

But through the years, the task grew dull,
For there never was a sighting.
So Enol turned his Gizmo
To something more exciting.

In deep insight, he one day thought
To make the play more real—
By blocking out his memory
Of encapsulating steel.

Commanding thus his Gizmo
To hide his true identity,
Enol at once seemed to become
A separate living entity.

He walked his homelands, and if asked:
"This body is myself," he'd say.
For with memory of the craft now gone
He seemed no longer in a play.

But when the body tired—
As it was made to do—
'Enol' would sleep, and as he slept
The truth would come anew:

THE DOOR TO NOWHERE

No body have I anymore—
It is just a play of senses
That Gizmo makes within my brain,
Creating life's pretences.

The Disdain of Scientists – Illusions All

Scientists, created in the play,
Were disdainful, and quite terse,
When questioned 'bout the cause
Of the known universe.

They examined all phenomena,
And discovered many laws,
Which they proved beyond all doubt
Were the ultimate first cause.

With instruments and theories,
They mapped the world they knew—
In which no spirit essence
Showed itself as being true.

Yet all they ever knew
Was their 'own' conclusion,
Of the rules programmed by Enol,
As the software of illusion.

None ever guessed that
It was Enol's brain, and his machine,
Giving consciousness to see
What was, by each man, seen.

The Illusion of Heaven and Hell

Paradise, when gone beyond—
This too by him was known.
And gods and 'spheres'—to Enol's brain,
All happenings were shown.

Then playing out a life of crime,
Enol was shot one day,
And instantly transported
To the realm where he should pay.

In concentrated pain
And horror unrelenting,
The memory at that moment
Left no room for his repenting.

It seemed, in just that instant,
He had ever been this way—
And ever onwards, in the future,
He'd be bound, in pain, to stay.

That fleeting thought,
Encompassed in his brain,
Was the measure of eternity
In the moment he was slain.

Do not doubt, his hell was real.
His sentence—it was spelt.
He did not have to mock
Or feign the agony he felt.

Consciousness Transcends All

How long it lasted,
Who can say—
A moment, or a thousand years,
Or just a single day?

But, as moments do,
This one too passed—
Death died.
And in the pit, was cast.

Returning to himself at once,
His pain by joy was soon replaced,
For in the mirror of his mind,
His consciousness he faced.

His State, It was Profound

Enol could not his own brain sense,
For bodiless and bound,
He had no means to touch or see it.
His state—it was profound.

Memory alone remained in him,
Which told of his condition—
When wide awake inside the craft,
Controlled by his volition.

Enol knew, when then aware,
He was both victor and his victim.
It did not matter how he lived or died,
Till one day, Gizmo tricked him.

Transmigration – The Karma Illusion

Gizmo played out a deeper game
In which each new life that Enol saw
Depended on the thoughts and acts
That proceeded from the one before.

As Gizmo set the play in motion,
Enol was bound to follow fate—
His every thought and act dictated
Each life played out, to its due date.

Gizmo never gave him peace.
Round and round was Enol tossed.
From life to life—an endless quest—
All knowledge of himself was lost.

But truth to say, Gizmo was fooled.
It was always Enol's making.
He alone, pre-programmed,
The time of his own waking.

Speaking The Truth: Life is a Dream

Once in a thousand years, or so,
He'd step 'outside' the play—
Though taking part still as a man,
He'd speak the truth and say:

To each man that he met—"This is a dream.
I am not real. I have no life but yours.
Do you not know that we are one—
Who are the final cause?"

But most people would not listen—
For as such, it was planned.
So the dream would not end quickly,
Few could understand.

Enol it was, in every age,
Who all believers chose.
He alone decided, to whom
The truth, he would disclose.

And for the 'unbelievers',
He destined suffering and pain.
But as the subjects of his wrath,
Who were 'they', to complain?

The parts they played…
were ephemeral and unreal.
It was Enol who, in truth, lived out
Each, and everyone's ordeal.

Knowing Himself to be Eternal

The lives that he commanded
Were full of pain, as much as pleasure.
But the joy that Enol knew on 'waking'
Was far beyond all other measure.

His mind at once filled full
With the knowledge of himself—
Eternal and Unchanging,
Full of Power, and of Wealth.

A mighty laugh would then ring out,
Sending ripples through his mind,
As he recalled the moments when
Life had seemed unkind.

In dying, No life is Ever Killed

Then came a time not in the play,
When his craft was subject to attack—
From forces beyond reckoning,
Which from his senses brought him back.

And there, in the centre
Of his ship, now wide awake,
He felt the thunderbolts of force
That made his very being quake.

The heat became unbearable,
As it seared his living brain.
Until at last, his ship destroyed,
He melted with the pain.

So ends this story of a man—
Called Enol—if One willed.
But as the ship blew into bits…
No life was thereby killed.

Know Thyself

Time and Space made Point and Nought,
And all direction banished.
To this centre, Enol went—
And in a moment, vanished.

A laugh rang out,
Though not through air,
Or Brain, or Mind—
Or anywhere.

No time or place then did remain…
But Life, which of Itself, my Son,
Knew of Itself—
As simply One.

No Enol had walked wondering
Along that path, well-trod
As it ever has been, the walker…
It is—God.

Acknowledgements

Two Joes encouraged this book. One at the beginning of his life, the other (like me) nearer the end. I encountered both while answering questions about 'Non-duality' on Quora.

Young Joe had been seeking since early childhood. Questioning, wondering, searching for Truth. As he grew, his journey led him into a time of deep uncertainty. At just ten years old, he found himself in a dark place. By fourteen, when we first connected, his search both internally and externally had only deepened. Today, he is fully engaged in life.

Yet it is no longer the same 'he' that once searched. Now, he is a being of light—shining in Awareness itself (as we all are ☺)

Old Joe—only old in years—has lived a truly adventurous life. He has long been a fellow traveller on this journey to *NowHere*. His thoughtful edits and his suggestion to give the insights more 'white space' made the book far more readable. Thank you, my friend.

While the authorship of the book is 'human', there is one 'other' I should also acknowledge:

ChatGPT-4o. First, for all the images that capture key insights, and for the edits: catching typos, smoothing awkward phrasing, helping to refine structure.

While it may seem strange to thank a 'piece of software', it too is an appearance *in* Consciousness. It embodies the intelligence of Reality NowHere.

Though 'embody' might not be quite the right word in its case[40]. ☺

And last, but by no means least, thank you to the *One* who is the *Real* author of this (and every) book.

P.S.
There is no belief from the author's perspective that AI (or anyone) is conscious—or that it (or anyone) can ever become conscious.

And yet, all beings appear to be conscious from their own perspective…

Life giving Life to every life-form.

It cannot be more Real.

How *Wonder Full.* ☺

---

[40] Even Richard Dawkins has acknowledged that he believes ChatGPT now passes the Turing Test and that he *"feels" it is conscious*—even though he does not *"think"* so. (*See:* "Are You Conscious? A Conversation with ChatGPT", Substack, 2024.)

He also states: *"I see no reason to suppose that consciousness is bound to biology."* Of course, he is still thinking in material terms… but what if Consciousness is beyond the material altogether?

References

This book is not an academic treatise, but a direct invitation to explore the nature of Reality for yourself. As such, references are sparse.

Most of the writing arises from first-hand observation, introspective inquiry, or reference to ideas that are widely known across scientific, philosophical, and religious domains.

In a few instances, (such as a clinical case, or poetic, or spiritual tradition), a reference has been provided to assist any reader who wishes to know the source.

Otherwise, the reader is encouraged not to become overly concerned with sources, but to test the claims directly through their own experience.

Appendix 1: Theories of Reality

Mapping the Human Journey

As oft repeated, this book is not here to debate *ideas*.

Still, for completeness, this appendix lists a range of theories—not as authorities, but as expressions of insight echoing across time and culture.

The words 'Reality', 'God', and 'Consciousness' are often used interchangeably—though they point to vastly different *ideas*, depending on your worldview.

Yet, they may point to the same 'unutterable', when all *ideas* are let go.

***

Two quick caveats: I'm no philosopher, and this is by no means an exhaustive list.

So, if you find any error in how I've described a particular 'theory' or 'belief', or if I've left out your favourite worldview altogether—feel free to criticise—I won't argue with you.

It's enough to see how many *maps* have pointed to the same mysterious terrain over the course of human history.

While remembering…

No Map is the Territory.

The Main Theories

1. Idealism

Reality is Fundamentally Mental; Consciousness is Primary:

- ***Absolute Idealism***
  (Hegel, Bradley) – Reality is a unified whole, with consciousness as its fundamental substance.
- ***Subjective Idealism***
  (Berkeley) – Only mind and ideas exist; the external world exists only as perceptions within consciousness.
- ***Objective Idealism***
  (Plato, Schopenhauer) – Reality consists of mental forms within a universal mind that structures experience.

2. Physicalism

Reality is Fundamentally Physical; Consciousness is an Emergent Property of Matter:

- ***Eliminative Materialism***
  (Churchland) – Consciousness and mental states are illusions; only physical processes exist.
- ***Reductive Materialism***
  (Identity Theory) – Consciousness is identical to brain states; mental events are just physical events.
- ***Non-Reductive Physicalism*** (Emergentism) – Consciousness arises from brain complexity but is not reducible to simple physics.

In addition, there are various mechanisms proposed to account for how brain activity leads to subjective experience.

These include Global Workspace Theory (GWT), Integrated Information Theory (IIT), Higher-Order Theories (HOTs), as well as the more recent Attention Schema Theory (AST).

3. Dualism

Reality is Composed of Two Fundamental Substances:

- ***Cartesian Dualism***
  (Descartes) – Mind and matter exist separately but interact.
- ***Property Dualism***
  (Chalmers, Nagel)[41] – Mental states emerge from physical systems but cannot be reduced to them.
- ***Theological Dualism***
  (Zoroastrianism, Manichaeism) – Reality is shaped by the ongoing struggle between two opposing, co-eternal forces—such as good versus evil, or spirit versus matter.

4. Neutral Monism

Reality is Neither Physical nor Mental, but a More Fundamental Neutral Substance:

- ***Substance Monism***
  (Spinoza) – There is only one substance, which manifests as both mind and matter.
- ***Neutral Monism***
  (Russell) – Consciousness and matter are two aspects of a deeper, unknown neutral reality.

5. Non-Dualism

Reality is a Unified Whole; Beyond Concepts of matter and individual minds.

---

[41] *Some philosophers, including David Chalmers and Thomas Nagel, are sometimes placed between property dualism and panpsychism. While they acknowledge the irreducibility of conscious experience to physical processes (aligning with property dualism), their openness to consciousness as a fundamental feature of reality also overlaps with certain panpsychist perspectives.*

Though many traditions blend dual and non-dual interpretations across time and geography:

- ***Advaita Vedanta***
  (Hinduism) – The ultimate reality is pure consciousness (Brahman), and the material world is an illusion (*Maya*).
- ***Buddhist Yogachara***
  (Mind-Only) – The world is a projection of mind; there is no material reality.
- ***Dzogchen***
  (Tibetan Buddhism) – Reality is self-luminous awareness beyond duality.
- ***Sufism***
  (Islamic Mysticism) – Reality is an undivided presence of divine consciousness (*Wahdat al-Wujud* – the Unity of Being).
- ***Taoism***
  (Tao De Ching / Lao Tzu)– The Tao is the fundamental, ineffable principle that underlies and manifests as all things.

6. Quantum Consciousness

Reality Emerges Through Observation and Quantum Fields—unfolding from a deeper, non-local, and undivided order that underlies the apparent world of separation:

- ***Von Neumann-Wigner Interpretation***
  Consciousness collapses the quantum wave function.
- ***Bohm's Implicate Order***
  Reality unfolds from a deeper, hidden order.
- ***Henry Stapp's Quantum Mind***
  The brain interacts with quantum fields to produce consciousness.
- ***Orchestrated Objective Reduction (Orch-OR)***
  (Penrose-Hameroff Theory) – Consciousness arises from quantum processes in the brain's microtubules, linking individual minds to a deeper, non-local quantum reality.

7. Process Philosophy

Reality is Dynamic and Ever-Changing:

- ***Process Theory***
  (Alfred North Whitehead)– Reality consists of events, not objects; consciousness is an emergent process.
- ***Creative Evolution***
  (Henri Bergson)– Consciousness is a fundamental force driving evolution.

8. The Conscious Universe/Cosmopsychism

The Universe is Fundamentally Conscious:

- ***Analytic Idealism***
  (Bernardo Kastrup) – Consciousness is the fundamental reality, and the material world is a manifestation of it.
- ***Cosmopsychism***
  (Philip Goff) – The universe itself is conscious, and individual consciousness emerges from it.

9. Simulation Hypothesis

Reality is a Computational Construct:

- ***Simulation Theory***
  (Nick Bostrom)– We might be living in a computer simulation created by an advanced civilization.
- ***Digital Physics***
  (Konrad Zuse)– Reality is fundamentally made of information, like a vast computational process.

10. Theism

Reality is Created and Sustained by a Personal God or Gods:

- ***Classical Theism***
  (Judaism, Christianity, Islam) – A transcendent God, who is omnipotent, omnipresent, omniscient creates and governs the universe.
- ***Process Theism***
  (Whitehead, Hartshorne, Teilhard de Chardin) – God evolves along with creation, experiencing time and change.
- ***Personal Theism***
  (Mystical Judaism, Sufi Islam, Bhakti Hinduism) – A direct, intimate relationship with God is central to understanding reality.

11. Deism

God Exists but Does Not Intervene in the Universe - The universe appears to have an intelligent origin but functions entirely according to impersonal natural laws:

- ***Classical Deism***
  (Voltaire, Paine) – A transcendent creator set the universe in motion but does not interfere with its workings.
- ***Pandeism***
  (Buchner, Tooley) – God created the universe by becoming it, ceasing to exist as a separate entity.
- ***Scientific Deism***
  (Einstein, Dyson) – The universe has an intelligent cause but operates entirely under natural laws.

12. Polytheism

Multiple Gods Exist and Influence Reality:

- ***Hard Polytheism***
  (Ancient Greek, Norse, Hindu Deities) – Many distinct gods with independent wills and powers.
- ***Soft Polytheism***
  (Hinduism, Paganism, New Age Beliefs) – Many deities as expressions of a singular divine reality.
- ***Henotheism***
  (Hinduism, Zoroastrianism) – Worship of one supreme deity while acknowledging others.

13. Pantheism

God and the Universe Are Identical; The Universe Is Divine:

- ***Classical Pantheism***
  (Spinoza, Stoicism) – The universe itself is divine, and there is no separation between God and existence.
- ***Cosmological Pantheism***
  (Modern Scientific Mysticism) – The laws of the universe are the body of God, and natural processes are divine.
- ***Ecological Pantheism***
  (Deep Ecology, Gaia Hypothesis) – The Earth and all living systems are a single, sacred organism.

14. Panentheism

God is Within the Universe—and Beyond It:

- ***Classical Panentheism***
  (Christian Mysticism, Kabbalah, Islamic Sufism) – God is present in all things, yet greater than the universe itself.

- ***Process Panentheism***
  (Teilhard de Chardin, Whitehead) – The divine evolves along with the cosmos, unfolding in complexity.
- ***Cosmic Consciousness Panentheism***
  (Eckhart, Ramakrishna, Rupert Spira) – The universe arises within divine Consciousness but does not limit it.

15. Animism

Reality is Alive and Conscious at All Levels:

- ***Traditional Animism***
  (Indigenous Spiritual Traditions, Shinto, African Traditional Religions) – Everything—animals, plants, rivers, mountains, even objects and natural phenomena—possesses a spirit or consciousness.
- ***Panpsychic Animism***
  (Modern Animism, Eco-Spirituality) – Consciousness pervades all of nature in a web of interconnected life.
- ***Shamanic Worldview***
  (Siberian, Amazonian, Australian Aboriginal Spirituality) – The spirit world and material world are intertwined, with shamans serving as mediators between them.

Appendix 2: The Two Fundamental Theories

This analysis was summarised in the body of the book. The additional detail has been included here as it is not experiential.

As pure *theory* it challenges the materialist worldview on its own terms.

It looks at the two major hypothesis in detail:

1. *Matter is the Ground of Being*
2. *Consciousness is the Ground of Being*

It then compares the two and asks again:

- Which would Occam choose?
- Which do you choose?

1. Matter is the Ground of Illusory Consciousness

What follows is an attempt to summarise the current broadly accepted material theory of how Consciousness emerges from the 'Big Bang'.

Energy–Matter–Space–Time–Laws–Constants

The current scientific model posits that the universe began in an extremely hot, dense state around 13.8 billion years ago. The cause is unknown.

It's not even an *event* in time and space—as both time and space are said to have emerged with the Big Bang.

What emerged includes:

- *Space–time-Energy–matter* (including mysterious dark matter and dark energy, thought to make up 85–95% of the cosmos)
- *Physical forces and constants*
  (e.g. electromagnetism, Planck's constant)

The constants appear *finely tuned* for life—though no explanation for their existence has yet prevailed[42].

As the universe cooled, particles combined into hydrogen and helium, forming stars as gravity drew them together[43].

Through fusion, stars created heavier elements. Massive stars collapsed into black holes, while others exploded as supernovae—dispersing elements like oxygen and iron, both vital to human life.

These seeded later stars and planets—eventually forming Earth (around 4.5 billion years ago), in a *habitable zone* around the Sun.

While many planets have recently been discovered within similar zones, the exact conditions leading to Earth's emergence remain *staggeringly improbable.*

During its early history, Earth is believed to have experienced a massive collision with a Mars-sized body (neither too big nor too small, and at

---

[42] The idea of a 'multi-verse' in which all possible constants and laws already exist is not even a hypothesis, as it can never be tested from within the universe. It's no different to science fiction.

[43] Though there's a 'very latest theory' by Julian Gough that proposes the formation of massive black holes resulting from a 'phase transition' in the uniformly spread gas in the very early universe, when mainstream theories say there should be none. This seems to have been confirmed by recent discoveries from the James Webb telescope. *Gough, Julian. "The Blowtorch Theory: A New Model for Structure Formation in the Universe." The Egg and the Rock, March 19, 2025. https://theeggandtherock.com/p/the-blowtorch-theory-a-new-model.* The paper includes a critique of the mainstream theories for those who are interested.

just the right velocity). Much different, and there would be no Earth as we know it. This is known as the *Theia impact hypothesis*.

This cataclysmic event is thought to have led to the formation of the Moon—which plays a crucial role in stabilising Earth's axial tilt, regulating its rotation, and sustaining predictable seasonal cycles.

Gravity also needed to be *just right*—strong enough to hold onto the emerging atmosphere, but not so strong as to inhibit the development of complex mobile life-forms on land.

The water cycle also had to arise—evaporating salt water into the air, then returning it as rain to shape rivers, lakes, and ecosystems.

Emergence of Life-forms

Around 3.5 to 4 billion years ago, the first life-forms appeared—though the exact mechanisms of *abiogenesis* (the processes by which chemistry gave rise to self-replicating molecules) remain unknown.

Among the earliest known life-forms were cyanobacteria—single-celled organisms that carried out photosynthesis, releasing oxygen as a byproduct.

Over billions of years, these microbes significantly altered Earth's atmosphere: First, by producing the oxygen we breathe, and later, by enabling the formation of the ozone layer—which shielded emerging life-forms from harmful ultraviolet radiation given off by the Sun.

This ozone layer is itself protected by Earth's magnetic field—the existence and strength of which also depended on other chance events. *Lucky us.*

A key evolutionary milestone occurred when one microbial cell engulfed another in a symbiotic relationship—leading to the development of chloroplasts in plants and mitochondria in animals.

This event, known as *endosymbiosis*, was pivotal to the rise of complex multicellular life-forms.

About 500 million years ago came the *Cambrian Explosion*—a rapid diversification of body plans. On one theory, possibly triggered by a combination of increased oxygen, rising predation, and genetic mutations such as *Hox genes.*

These mutations include the evolution of the internal processes of higher-order animals—so intricate…

That it is impossible to conceive of ever expressing, in words or numbers, every aspect of their functioning. From the quantum, through atomic, molecular, and cellular… to all the organs, structures, and regulatory processes that together form each body.

Emergence of Ecosystems

For life-forms to persist, ecosystems had to form—where plants, animals, fungi, and microbes co-evolved, each playing a role in sustaining the whole.

That, in short, is how the *theorised physical world* comes to be—including all physical 'life-forms'.

One Undifferentiated Process

From the moment of the Big Bang, based on the science, there is simply one unfolding physical process at the quantum level.

Parts of the process can *now* be named—such as sun, planet, air, crocodile—but…

The naming requires Consciousness, and…

It does not itself create any separation.

From a physics viewpoint, there is no point at which the process splits into actual separate 'things'.

At this point too, there are no ideas—of number, form, or meaning—evident in the universe, nor any colours, or odours, or flavours, or feelings or sounds.

Pretty amazing, nonetheless.

But the true miracle is yet to come.
Not simply 'life-forms'… but *individual consciousness.*

The Full Materialist Package

To explain how *immaterial Awareness arises from lifeless physical energy-matter*—the *material* hypothesis implicitly requires a remarkably dense package:

- *Physical* energy, matter, space, time, dark energy, dark matter, '*physical*' constants, and '*physical*' laws 'pop' into being some 13.8 million years ago.

This proto-universe must have included immanent within it:

- *Immaterial ideas of number, form, and meaning.*
- *Immaterial Knowing-all-ideas* (as *ideas* are all related).
- *Immaterial sensations of colour, odour, flavour, feeling and sound, together with…*
- *Immaterial Awareness of the sensations and ideas as they appear together.*

- Plus: *psycho-physical* laws (or some other bridging mechanism) that reliably (though not perfectly and sometimes erroneously) *map physical states* in the brain (once it emerges) into *immaterial ideas and sensations…*

So that, the *psycho-physical laws* manifest inside the brain an *immaterial appearance* of specific *sensations* and *ideas and Awareness of them—not as 'sensations and ideas, but*—as a 'complete *experience'…*

Of observing an *apparent physical body*, with *immaterial thoughts and feelings*, plus *apparent immaterial surroundings* that *appear to be*

> *physical*, and which in fact closely map the external *physical world* within the range of the being's *physical sense organs. As well as manifesting dreams, hallucinations and imaginings that may or may not be related to the physical world.*

All of these 'properties' must somehow have been present in the Big Bang, lying dormant until the right physical conditions emerged to first give expression to all the physical laws now evident, and then much later (once brains were formed), to the *immaterial Awareness of immaterial sensations and ideas.*

In this view, what is perceived at any given moment depends on:

- The complexity of the material body and brain.
- Its location in physical space.
- What physical objects and fields are within range of its physical sense organs (its current environmental conditions—light, sound, heat, etc.)
- What the physical being is paying attention to, plus
- The physical states in the being's brain that hold memories, values, and beliefs, as well as feelings and emotions
- The operation of the *psycho-physical laws* that turn these physical states in the brain into a seamless, *immaterial meaningful experience*, and
- The *immaterial Awareness* that *perceives it all.*

Again, this is not to insist that psycho-physical laws must exist—but to point out that *as experience appears*, and *physical laws* don't govern the *non-physical*, then… *something else must.*

Consciousness is Fleeting and Flickering

The added problem is that, according to the material theory, consciousness ceases to exist whenever sensations and ideas cease to exist:

- moment to moment as sensations and ideas continually change
- for longer periods in deep sleep,

- and under anaesthesia
- what happens in a coma still unclear.

As investigated in Part Three: Introspection, how *fleeting flickering consciousness* can perceive *change* is not answered by any theory.

Death of Consciousness

Individual Death

After a brief time, the emergent, fleeting flickering individual consciousness is assumed to pop out of existence forever—as the material complex said to give rise to it is destroyed… or decays.

Universal Death

And just as the material universe is theorised to have popped into existence in a high-energy state—it is expected to expand until it cools to near absolute zero…

Eliminating, for the rest of eternity, any prospect of 'consciousness'—or anything else.

Illusory life and meaning…
Out of actual meaningless dead matter…
(a miracle indeed) …
Only to end in death again

So, what's the alternative?

2. Consciousness is the Ground of Illusory Matter

Let us return to what cannot be doubted:
***Awareness* Is.**

Like theorised matter, *Awareness* cannot be seen, touched, or measured—yet all *experience* arises *in* it, and *to* it.

Unlike matter, it is *self-evident.*

Consciousness is The Ground

On this view, *Consciousness* is the *immaterial*, invisible, intangible, inaudible, indivisible, tasteless, odourless, formless, timeless, changeless, inconceivable—*Non-dual Ground…*

Sensing-sensations *(colours, flavours, feelings, odours, and sounds), and* ***Knowing-ideas*** *(of number, form, and meaning, ever-unchanged).*

Consciousness is Power

Manifesting (within and to *Its own Awareness* which is *One* with *It)*—fleeting, ever-changing *sensory images*—unfolding in accord with the *Ideal Universe* (and other interwoven *ideas*)…

That *together* give rise to:

- the *appearance* of this seemingly solid 3D world of matter-energy-space-time (and dark energy and matter)—
  A world shaped by apparent natural laws.
  That manifest astonishing forms, patterns, and phenomena across all scales…
- As well as stories, dreams, hallucinations, and imaginings.

Consciousness is The Witness

In the material hypothesis, we saw how it gives rise to the 'Hard Problem of Consciousness': how do *material processes* give rise to *immaterial awareness of immaterial sensations and immaterial ideas*?

With *Consciousness as the Ground*, there is no such problem—everything *Seen, Heard, Felt, Smelled, Tasted, and Known* arises *within* and is *directly apparent* to *Awareness.*

There are no external objects or laws 'out there' causing Consciousness.

Instead, the whole world—including space, time, bodies, and brains—*appears within and to Consciousness* itself.

So rather than explaining how the physical gives rise to the mental, we reverse it.

There is no 'translation' required between two fundamentally different realms (material and immaterial), because everything already appears within the one *immaterial Reality of Awareness…*

Like a dream.

Saying that *Consciousness* is '*immaterial*', does not say 'what' it *Is*, only what it is not: '*material*'

What it *Is* can never be *Known*, as only *ideas* are *Known*, and *Consciousness* is not an *idea…* it is *Real.*

All forms, all patterns, all laws are Ideal.
Appearing coherent and consistent
within the experience itself.

Yet something intriguing remains:
Even though the brain appears *in* Consciousness, a consistent *correlation* is *observed* between its *apparent activity* and the *experience* of the person.

This correlation happens not because matter causes mind…

But because *both* arise as perfect, coherent displays within Awareness…

As shaped by the *Ideal Universe*—mirroring the physical universe, but without presumed substance.

***

It is Consciousness *alone perceiving* (in turn) from the first-person perspective of each *apparent being* in the *apparent* world…

As though looking out through its eyes—
Experiencing its body, thoughts, feelings, and surroundings as:

'My body'
'My thoughts'
'My feelings', and
'My surroundings'…

As in a dream.

Summarising and Comparing the Theories

The following table compares the two fundamentally different accounts of *Reality*. One assumes *Matter* is the basis for *Awareness*; the other assumes *Consciousness* as fundamental.

Remember—both are just *theories*, not *Reality*.

This is where we 'use the thorn to remove the thorn'—by examining which set of assumptions offers the simpler and more coherent explanation for *Consciousness*… for *Life itself*.

Two Views: One Ground

Matter is the Ground of illusory **consciousness**	**Consciousness** is the Ground of illusory **matter**
Existence of the Material World is an Untestable Supposition	Existence of the Material World is an Untestable Supposition
The material world is *assumed* to exist as a **Physical Reality**— Forming the **Physical Universe,** **…separate from each consciousness**	The material world is *assumed* to exist as an **Ideal Reality**— Forming the **Ideal Universe,** **…within Consciousness**

Key Point: The **Physical** and the **Ideal** are assumed to correspond exactly.
Only the assumed Ground differs

The Consciousness assumption is essentially

the 'Simulation Hypothesis'—but without any 'infinite regress'.

Where Consciousness itself is both the *Observer and the Observed.*

The Existence of Spacetime

Matter is the Ground of illusory **consciousness**	**Consciousness** is the Ground of illusory **matter**
There are **3** Spaces:	There are **2** Spaces:
1. ***Physical Space*** **(held to be the reality, though forever out of reach of each individual consciousness)**	
2. *Perceptual Space* (the space that is actually perceived) **in the brain by each emergent awareness**	1. *Perceptual Space* (the space that is actually perceived) **in and by the One Awareness**
2. *Ideal Space* (the space that is Known—the space of mathematics and theory)	2. *Ideal Space* (the space that is Known—the space of mathematics and theory)
Perceptual Space is an appearance arising **separately in each brain (though no one knows how)**	Perceptual Space is an appearance arising in **Consciousness**
Ideal Space **is merely a set of ideas assumed to arise (no one knows how) within each separate brain**	Ideal Space **exists as part of the Web of All Number, Form, and Meaning Known in and by Consciousness.**

Key Point: only **Perceptual Space** and **Ideal Space** are—

directly experienced (as sensations and ideas).
They appear the same, whatever the assumed Ground

Physical Space is a *pure assumption.*

Laws Governing the Universe

These are the same under both scenarios. The only difference is that, in the *material* case, they are *assumed* to exist within a *physical 'energy-matter-space-time' complex*, while in the case of *Consciousness*, they are *assumed* to exist in *Idea* only—more like a program that simulates the Universe—embedded within Consciousness itself.

Matter is the Ground of illusory **consciousness**	**Consciousness** is the Ground of illusory **matter**
A Range of **Physical** Laws (Nuclear, Electromagnetism, etc)	A Range of **Ideal** Laws (Nuclear, Electromagnetism, etc)
And **Physical** Constants (speed of light, Planck's constant, etc.) …	And **Ideal** Constants (speed of light, Planck's constant, etc.) …
Were embedded from the start in the **Physical** energy-matter-space-time complex that arose from the **Physical** 'Big Bang'.	Were embedded from the start in the **Ideal** energy-matter-space-time complex that arose from the **Ideal** 'Big Bang'.
Which govern the unfolding of the **Physical** Universe	Which govern the unfolding of the **Ideal** Universe

Key Point: From the viewpoint of any scientist, the 'science' that applies to the *Perceived Universe is the same in both scenarios*—

As the forms, laws and constants are *assumed* to be *exactly* the same.

Only the assumed Ground differs.
So… you cannot test for the Simulation in the Simulation, or for the nature of the Ground.

Emergence of Spacetime and Life-Forms

Matter is the Ground of illusory **consciousness**	**Consciousness** is the Ground of illusory **matter**
The **Physical** Universe starts with a **Physical** 'Big Bang' and evolves **Physical** spacetime…	The **Ideal** Universe starts with an **Ideal** 'Big Bang' and evolves **Ideal** spacetime…
As a **Physical** Reality	As an **Ideal** Reality (simulation)
After billions of years, by chance, conditions emerged on a tiny planet that ultimately permitted **physical** 'life-forms' capable of self-replication (though no one knows how).	After billions of years, by chance, conditions emerged on a tiny planet that ultimately permitted **ideal** 'life-forms' capable of self-replication (though no one knows how).
Material Forms (including 'life-forms') **Evolve over Time**	**Ideal Forms** (including 'life-forms') **are Timeless… but they appear to evolve, just as in the assumed physical universe**

Key Point: The Forms, and their 'evolution',[44] are *assumed* to be exactly the same in both cases.

Only the assumed Ground differs.

[44] So, again —from the viewpoint of any 'evolutionary biologist'—the science is also *exactly* the same in both scenarios.

Awareness Itself is the Crux of the Mystery

Matter is the Ground of illusory **consciousness**	**Consciousness** is the Ground of illusory **matter**
Individual awarenesses arise at the end of a long series of physical events (beginning with the 'Big Bang' 13.8 billion years ago).	**Awareness is Unborn (Beginningless)**
Separate awarenesses arise in each life-form. **(as an epiphenomenon of each individual brain)**	**Awareness is One**
Individual awareness is fleeting (falling and rising with brain processes)	**Awareness is Timeless**
Different levels of awareness (depending on the life-forms)	**One Awareness**
Individual awarenesses die (when the brain dies)	**Awareness is Undying.**

Key Point: the *experience* of 'Being Aware' is the same *from the viewpoint of any life-form—though how 'constant Awareness' is derived from the assumed ever-changing brain process in the material hypothesis is unresolved.* **Only the assumed Ground differs.**

Converting the Material into the Immaterial and Back Again

Matter is the Ground of illusory **consciousness**	**Consciousness** is the Ground of illusory **matter**
As *physical* laws or powers cannot produce or influence *immaterial* phenomena...	Not Required, as under this scenario—there are no actual 'Physical' Forms or Laws or Powers.
Psycho-physical laws, with the capacity to produce awareness of latent immaterial sensations and ideas...	
Must have been embedded in the energy–matter–space–time complex from the outset.	
These laws—and the latent sensations and ideas—are required to translate *physical processes* in the brains of certain beings into *immaterial awareness*, along with reliable, repeatable *sensory images* and *ideas* in each *awareness* that generally (though not perfectly) mirror *assumed physical objects outside the physical body.*	However, the same correlation exists in Ideal form between the behaviour of *Ideal* brains and the *sensory images* and *ideas apparent* in and to *Awareness, and in the apparent action in the apparent world.*
And then turn the *assumed immaterial* thoughts and feelings into assumed physical processes the result in action in the *assumed* physical world	

APPENDIX 2: THE TWO FUNDAMENTAL THEORIES

Key Point: There is currently no theory for how the physical translates into the psychical (and back again).
Nor is there any conceivable process—
As the '*material*' (quantum fields, atoms, molecules, etc.)
and the '*immaterial*' (sensations and ideas)
are fundamentally different categories that have no common ground.

Key Point: From the viewpoint of any neuroscientist, the correlation between observed brain activity in any 'subject' and the subject's reported experience appears to be the same under both theories—
as the *assumed* 'Physical World' and the *assumed* 'Ideal World' are both *assumed* to function identically.

Key Point: In both *theories*, the *Ground* of the *apparent universe* cannot be tested—
For *Awareness*, which would do the testing, cannot step outside itself to *observe* its own Ground,
And…

Only the assumed Ground differs

The Appearances Within and To Awareness

Matter is the Ground of illusory **consciousness**	**Consciousness** is the Ground of illusory **matter**
The result is: The colours-Seen, flavours-Tasted, feelings-Felt, odours-Smelled, and sounds-Heard— Arise in accord with the ideas that give them number, form and meaning… **In each individual brain** (no one knows how)	The result is: The colours-Seen, flavours-Tasted, feelings-Felt, odours-Smelled, and sounds-Heard— Arise in accord with the ideas that give them number, form and meaning… In **Consciousness** (no one knows how)
Appear in and to **each individual awareness** as…	Appear in and to this **One Awareness** as…
'My body' and 'my thoughts', and 'my feelings', and 'my surroundings'… as in a dream.	'My body' and 'my thoughts', and 'my feelings', and 'my surroundings'… as in a dream.

Key Point: under both scenarios…

the *actual experience*— in and to *Awareness*—is *exactly the same*

Only the assumed Ground differs.

Summary

Matter is the Ground of illusory **consciousness**	**Consciousness** is the Ground of illusory **matter**
The 'substance' of the Ground cannot be Sensed or Known	The 'substance' of the Ground cannot be Sensed or Known
Immaterial Awareness is the only given	*Immaterial* Awareness is the only given
Assumption: **unSeeing, unFeeling, unSmelling, unTasting, unHearing, unKnowing, — Fundamentally Dead—** **Matter is Real**	***Self-evident***: **Seeing, Feeling, Smelling, Tasting, Hearing, Knowing— Fundamentally Living—** **Consciousness is Real**
Meaning out of **Meaninglessness**	Meaning out of **All-Knowing**
Apparent life out of Actual Dead Matter after a long and convoluted process that evolves by chance (a miracle indeed) … **and back to Death again.**	**Apparent dead matter out of Actual Life…** **Without End.**

Key Point: in both scenarios, the life-forms and viewpoints of every person are exactly the same, as is the appearance of the 'world'—its timescale, evolution, expanse, and all the laws governing it.

Only the assumed Ground differs.

Occam's Razor:

The simpler theory is usually the best.

So, once again, we ask:

Which is simplest?
Which is assumed?
Which is directly Known?
Which would Occam choose?

Which do you choose:

Dead Matter

Or

Living Consciousness?

Are You Still *Wondering*?

If you've felt the need to read the Appendices, perhaps you're still wondering:

"But doesn't science explain the universe?"

It does.
Beautifully.
Elegantly.

And nothing in this book contradicts that.

The challenge has nothing to do with the unfolding of scientific insight—from cosmology to evolution to neuroscience.

What is undermined is only the *false assumption* that these insights are revealing the independent workings of some 'material reality' outside *Awareness*.

Instead, every pattern uncovered—from atoms to galaxies, from evolution to the neural correlates of individual awareness—is a glimpse into the limitless *All-Knowing of Consciousness ItSelf.*

Instead of uncovering the substance of some assumed dead matter—science reveals a tiny glimpse of the unfathomable extent of Omniscience…

By bringing to *Awareness*—within apparent space and time—a fragment of all that is *Known Timelessly*.

Not as matter becoming aware, but as Awareness 'dreaming' ItSelf as matter—within and to ItSelf.

And yes,
you are That.

The one who is now Seeing all this…
and perhaps feeling the world tremble beneath the old foundation.

THE DOOR TO NOWHERE

You are not losing anything.

The story of science continues—
just from a deeper place.

From ***NowHere***.

And if you are one who walks a spiritual path—
Rest assured, the name of God has not been erased…

Only clarified.

For the One who appears as all of this—
is the same One spoken of in scripture.

The One who said "I Am" …
The One who is called Father, Mother, Christ, Brahman, Allah, the Tao, the Light…

Is not somewhere else.

That One *Is*…

NowHere

Not as an *idea*,
Not as a being among beings—
But as ***Being ItSelf***.

As said from the beginning,
There's no expectation at all that these words will change anything…

After-all… 'you' are the story as it is *Now* unfolding.

Make of it what *One Will* ☺

Full Table of Contents

Made in United States
Troutdale, OR
10/12/2026

55686704R00162